PARASITES AND PARASITIC DISEASES

ENTAMOEBA

SPECIES, CLASSIFICATION AND BIOLOGY

PARASITES AND PARASITIC DISEASES

Additional books and e-books in this series can be found on Nova's website under the Series tab.

PARASITES AND PARASITIC DISEASES

ENTAMOEBA

SPECIES, CLASSIFICATION AND BIOLOGY

THOMAS L. JOHNSON
EDITOR

Library of Congress Cataloging-in-Publication Data

Names: Johnson, Thomas L. (Nova Publishers editor) editor.
Title: *Entamoeba*: species, classification and biology / Thomas L. Johnson, editor.
Description: New York: Nova Science Publishers, [2021] | Series: Parasites and parasitic diseases | Includes bibliographical references and index. | Summary: "*Entamoeba*: Species, Classification and Biology outlines the current knowledge about the global epidemiology, clinical manifestations, pathogenesis, available diagnostic tools and management of pathogenic *Entamoeba* species in humans. The authors provide an overview of the various species belonging to the genus *Entamoeba*, including *Entamoeba histolytica*, *Entamoeba dispar*, *Entamoeba* gingivalis, *Entamoeba* coli, *Entamoeba moshkovskii*, *Entamoeba* invadens, *Entamoeba* polecki, *Entamoeba hartmanni*, *Entamoeba* suis, *Entamoeba* nuttalli, *Entamoeba bangladeshi*, *Entamoeba* struthionis and *Entamoeba* muris. The general pathogenic factors of *Entamoeba histolytica* are described, particularly focusing on how these factors participate in establishing infection. Additionally, a review of studies centered on the morphological, biochemical and genetic changes during programmed cell death of *Entamoeba histolytica* is provided. Current knowledge on the identification, characterisation, structure, and function of the enzymes of serine biosynthesis is also summarized"-- Provided by publisher. Identifiers: LCCN 2020043022 (print) | LCCN 2020043023 (ebook) | ISBN 9781536185065 (paperback) | ISBN 9781536187441 (adobe pdf) Subjects: LCSH: *Entamoeba*. Classification: LCC QL368.A5 E58 2021 (print) | LCC QL368.A5 (ebook) | DDC 579.4/32--dc23
LC record available at https://lccn.loc.gov/2020043022
LC ebook record available at https://lccn.loc.gov/2020043023

Published by Nova Science Publishers, Inc. † New York

CONTENTS

PREFACE

Entamoeba: Species, Classification and Biology outlines the current knowledge about the global epidemiology, clinical manifestations, pathogenesis, available diagnostic tools and management of pathogenic *Entamoeba* species in humans.

The authors provide an overview of the various species belonging to the genus *Entamoeba*, including *Entamoeba histolytica*, *Entamoeba dispar, Entamoeba gingivalis*, *Entamoeba coli*, *Entamoeba moshkovskii, Entamoeba invadens*, *Entamoeba polecki*, *Entamoeba hartmanni, Entamoeba suis*, *Entamoeba nuttalli*, *Entamoeba bangladeshi*, *Entamoeba struthionis* and *Entamoeba muris*.

The general pathogenic factors of *Entamoeba histolytica* are described, particularly focusing on how these factors participate in establishing infection.

Additionally, a review of studies centered on the morphological, biochemical and genetic changes during programmed cell death of *Entamoeba histolytica* is provided.

Current knowledge on the identification, characterisation, structure, and function of the enzymes of serine biosynthesis is also summarized.

Chapter 1 - The genus *Entamoeba* comprises of ubiquitous parasites belonging to the phylum Amoebozoa. In this genus, *Entamoeba histolytica (E. histolytica)*, a known pathogen, is the causative agent of amoebiasis which is a leading cause of diarrhea worldwide. This species had long been heralded as the only pathogenic species in man. However, with the development of molecular methods for identification, other species are also being reported with varying pathogenic potential. Species like *Entamoeba dispar (E. dispar)* and *Entamoeba moshkovskii (E. moshkovskii),* both morphologically identical to *E. histolytica* and previously considered as commensals, are being reported sporadically in association with probable intestinal infections. Recently, *Entamoeba bangladeshi (E. bangladeshi)* a novel species has also been associated with the intestinal infections in children. Large number of infections of *Entamoeba* spp. remains asymptomatic. Others manifest as diarrhea in majority of the cases while *E. histolytica* can cause further invasive infections in form of colitis and extra intestinal abscesses in different vital organs. Several host and parasite factors decide the final outcome in such infections. The epidemiology of *Entamoeba* spp. has been studied worldwide. Though limited to subtropical and tropical regions in resource limited countries, cases of imported amoebiasis has been a growing burden even in developed countries. Diagnosis of these infections had been largely unreliable due to inability of microscopy to distinguish between the species until the development of molecular methods. Nitroimidazoles remains the drug of choice for treatment of *Entamoeba* infections often in combination with luminal cysticidal agents. However, occasional reports of clinical non-responsiveness have prioritized the need for development of a new drug. Amoebiasis, being a global health problem, this parasite requires renewed attention for better understanding and formulation of strategies for prevention and control. This chapter outlines the current knowledge about the global epidemiology, clinical manifestations, pathogenesis, available diagnostic tools and management of pathogenic *Entamoeba* species in humans.

Chapter 2 - The genus *Entamoeba* comprises engrossing protists, which are found in humans, nonhuman primates, other vertebrates and invertebrates. *Entamoeba histolytica*, *Entamoeba dispar*, *Entamoeba gingivalis*, *Entamoeba coli*, *Entamoeba moshkovskii*, *Entamoeba invadens*, *Entamoeba polecki*, *Entamoeba hartmanni*, *Entamoeba suis*, *Entamoeba nuttalli, Entamoeba bangladeshi, Entamoeba struthionis* and *Entamoeba muris* are important organisms of the genus. Of these, a few species are harmless, while a few are pathogenic. *Entamoeba gingavalis* was the first parasitic amoeba reported in humans, later on some other species have been found in human, of these, *Entamoeba histolytica* is most fatal species that affects more than 10% of the world's population and causes amoebiasis, which is an emerging parasitic complication in the human immunodeficiency virus (HIV)-infected patients and is responsible for approximately 100,000 fatalities annually, making it the second leading cause of death due to parasitic disease. Moreover, the species which are morphologically indistinguishable from *E. histolytica* are of immense importance because these may be confused with *E. histolytica* in diagnostic investigations. In order to control infectious agent, sound knowledge of taxonomy, prevalence, morphology and mode of transmission of the infectious agent is extremely essential, therefore, in this context, the present chapter provides the details of various species belonging to the genus *Entamoeba*.

Chapter 3 - *Entamoeba histolytica* is a protozoan parasite with high prevalence in developing countries that causes amoebiasis. This disease affects the intestine and the liver; and it is the third leading cause of human deaths among parasite infections. Establishing an amoebic infection involves several crucial steps, such as degradation of the mucosal layer, adherence to the intestinal epithelium, invasion into the tissues, and dissemination to other organs. Each one of these infection steps depends on pathogenic factors that allow invasion and destruction of the host tissues. For invasion, the main pathogenic factors have been described and include galactose (Gal) and N-acetyl-D-galactosamine (GalNAc) lectins, glycosidases, proteolytic enzymes, and parasite-produced cytokines. For tissue destruction the amoeba uses two main mechanisms, contact-dependent cytolysis, and trogocytosis, in which amoebas nibble pieces of live cells. Once amoebas invade the tissues, they have to survive in the host by evading and confronting the immune system. Antibody production is induced by amoebas, but immunoglobulins (Ig) G do not confer protection. IgA antibodies are a major component of the human intestinal defense mechanism, and they have been shown to confer some protection in animal models of amoebiasis. Also, interferon-gamma (IFN-γ) production is involved in clearance of infection, probably via activation of leukocytes. In addition, *E. histolytica* infection of the intestine or liver is associated with a strong inflammation characterized by a large number of infiltrating neutrophils. Consequently, it has been suggested that neutrophils play a protective role in amoebiasis. However, some reports indicate that trophozoites making direct contact with neutrophils provoke lysis of these leukocytes, resulting in the release of their lytic enzymes, which in turn provoke tissue damage. Therefore, the role of neutrophils in this parasitic infection remains controversial. Recently, it was found that *E. histolytica* trophozoites are capable of inducing neutrophil extracellular traps (NET) formation, while the non- pathogenic *Entamoeba dispar*, which cannot be distinguished from *E. histolytica* by microscopic analysis, was not able to induce NET formation. Thus, NET formation seems to be a novel defense mechanism against amoebiasis. In this chapter, the authors will describe, the general pathogenic factors of *E. histolytica,* and how these factors participate in establishing infection. Also, the authors will summarize the latest knowledge on immune response and immune evasion during amebiasis.

Chapter 4 - Amoebiasis is caused by the protozoan *Entamoeba histolytica* and primarily affects developing countries. Because effective vaccines against this parasite have not been developed, the treatment of amebiasis is primarily pharmacological. However, there are reports of treatment failure and the generation of drug-resistant clones *in vitro* and *in vivo.* For this reason, the study of programmed cell death (PCD) of this parasite is important since knowledge of its underlying molecular mechanism may offer novel therapeutic tools to combat the parasite. Advances in the understanding of the PCD of *E. histolytica* have demonstrated that PCD is induced by nitric oxide species, the aminoglycoside G418, hydrogen peroxide, natural compounds and host-parasite interactions. Morphological, biochemical and genetic apoptotic signals have been observed, including cell shrinkage with an increased number of vacuoles, nuclear condensation, chromatin fragmentation, preservation of the trophozoite cell-membrane integrity, externalization of phosphatidylserine (PS), overproduction of reactive oxygen species (ROS), increases in cytosolic calcium, and decreases in intracellular potassium, ATP levels and pH. The diversity of regulatory signals suggests a calpain-dependent pathway and that a caspase-dependent mechanism could be present in *E. histolytica.* Herein, the authors provide an overview of studies centered on the morphological, biochemical and genetic changes during PCD of *E. histolytica.*

Chapter 5 - The growth, survival, and virulence of the microaerophilic protozoan *Entamoeba histolytica* are correlated with its redox power and expression of its anti-oxidative enzymes. Thus, the ability to subvert the toxicity of oxidative stress is of central importance for the establishment of successful infection. Cysteine is the major thiol present, and one of the key molecules that help in redox defence of the parasite *E. histolytica.* L-serine is the only substrate for the *de novo* cysteine biosynthesis. D-phosphoglycerate dehydrogenase (PGDH), 3-phosphoserine aminotransferase (PSAT), and phosphoserine phosphatase (PSP) are the three enzymes catalysing the serine synthesis in the so-called phosphorylated pathway from the glycolytic intermediate 3-phosphoglyceric acid (3-PGA). PGDH catalyses the first committed step of this pathway, and EhPGDH was the first structure to be reported from the Type IIIK PGDH family. The rate-limiting step of serine biosynthesis is carried out by PSP, and *E. histolytica* possesses a novel metal independent PSP, in contrast to its human host. This chapter aims to summarise the authors' current knowledge on the identification, characterisation, structure, and function of the enzymes of serine biosynthesis in *E. histolytica.* The recent findings also reveal that the serine biosynthesis is capable of counteracting the increased cysteine synthesis demand under oxidative stress conditions.

In: *Entamoeba*
Editor: Thomas L. Johnson
ISBN: 978-1-53618-506-5

Chapter 1

PATHOGENIC INTESTINAL *ENTAMOEBA* SPECIES IN HUMANS: EPIDEMIOLOGY, PATHOGENESIS AND MANAGEMENT

Aradhana Singh and Tuhina Banerjee*
Department of Microbiology, Institute of Medical Sciences, Banaras Hindu University, Varanasi, India

ABSTRACT

The genus *Entamoeba* comprises of ubiquitous parasites belonging to the phylum Amoebozoa. In this genus, *Entamoeba histolytica (E. histolytica)*, a known pathogen, is the causative agent of amoebiasis which is a leading cause of diarrhea worldwide. This species had long been heralded as the only pathogenic species in man. However, with the development of molecular methods for identification, other species are also being reported with varying pathogenic potential. Species like *Entamoeba dispar (E. dispar)* and *Entamoeba moshkovskii (E. moshkovskii),* both morphologically identical to *E. histolytica* and previously considered as commensals, are being reported sporadically in association with probable intestinal infections. Recently, *Entamoeba*

* Corresponding Author's E-mail: drtuhina@yahoo.com.

bangladeshi (E. bangladeshi) a novel species has also been associated with the intestinal infections in children. Large number of infections of *Entamoeba* spp. remains asymptomatic. Others manifest as diarrhea in majority of the cases while *E. histolytica* can cause further invasive infections in form of colitis and extra intestinal abscesses in different vital organs. Several host and parasite factors decide the final outcome in such infections. The epidemiology of *Entamoeba* spp. has been studied worldwide. Though limited to subtropical and tropical regions in resource limited countries, cases of imported amoebiasis has been a growing burden even in developed countries. Diagnosis of these infections had been largely unreliable due to inability of microscopy to distinguish between the species until the development of molecular methods. Nitroimidazoles remains the drug of choice for treatment of *Entamoeba* infections often in combination with luminal cysticidal agents. However, occasional reports of clinical non-responsiveness have prioritized the need for development of a new drug. Amoebiasis, being a global health problem, this parasite requires renewed attention for better understanding and formulation of strategies for prevention and control. This chapter outlines the current knowledge about the global epidemiology, clinical manifestations, pathogenesis, available diagnostic tools and management of pathogenic *Entamoeba* species in humans.

INTRODUCTION

Entamoebae were considered to be most primitive extant eukaryotic group, Archezoa [1]. The *Entamoeba* genus comprises of a group of anaerobic, unicellular, parasitic organisms which contributes to heavy burden of diarrhea especially in developing countries with low socio-economic background and poor sanitation [2].

Entamoeba histolytica (E. histolytica) is the causative agent of amoebiasis and has been associated with considerable mortality and morbidity among children [3]. *E. histolytica* infections accounted for approximately 55,000 deaths in 2010 and 2.237 million disability adjusted life years (DALYs) i.e., the sum total of years of life lived with disability and years of life lost [4]. Because of this reason, *E. histolytica* remains the focus of majority of studies. However, recent evidences have shown that *Entamoeba moshkovskii* (*E. moshkovskii*) and *Entamoeba bangladeshi* (*E. bangladeshi)* are also associated with diseases in humans [5-6].

Entamoeba dispar (*E. dispar)* is the new name given to what was previously called as non-pathogenic or non-invasive *E. histolytica*. However, reports of association of *E. dispar* with intestinal symptoms have raised doubt on its pathogenicity [7]. Majority of the *Entamoeba* infections remain asymptomatic and only 10% develop clinical symptoms ranging from intestinal colitis to dysentery to extra intestinal lesions of lungs, liver, heart and brain [8]. The reason behind such partial virulence remains unknown, though it is believed to be an interplay of host factors, parasite virulence factors and surrounding micro-environment [9].

Despite the presence of specific and sensitive molecular techniques, microscopy because of its affordability and accessibility remains the diagnostic tool for the *Entamoeba* infections [10]. Microscopic examination cannot differentiate other related species from *E. histolytica*. Consequently, World Health Organization (WHO) has recommended the use of advanced techniques in the diagnosis of *Entamoeba* infection [11]. New approaches based on antigen and antibody detection in the clinical samples has resulted in better identification of *Entamoeba* species in the developing countries [12-13]. Additionally, advanced molecular techniques such as real time polymerase chain reaction (PCR), microarray has redefined the distribution of the *Entamoeba* genus in many geographical regions [10, 14-16].

Nitroimidazoles remains the drug of choice for *Entamoeba* infections from last many decades along with some luminal agents [17]. Till date, there is no vaccine available for *Entamoeba* infections and the concerns over the toxicity and development of resistance against nitroimidazoles is also emerging [18]. Sporadic studies of recurrence of *Entamoeba* infections after appropriate therapy have been reported [19-21]. Since last 60 years, there has been no drug development against amoebiasis, despite priorities set by National Institute of Allergy and Infectious Diseases (NIAID) for drug development for treatment of category B pathogens [17]. *Entamoeba* infections being a global health problem, especially in children, requires proper attention for its control and management. With this background, this chapter covers the different aspects on the pathogenic *Entamoeba* species in humans.

ENTAMOEBA SPECIES INFECTING HUMANS

Over 145 years ago, in 1875, Fedor Losch described motile amoeba in the stool sample of laborer suffering from dysentery [22, 23]. However, he concluded that dysentery was caused by bacteria and the presence of amoeba just added to the inflammation. Subsequently in next 16 years, many authors found concomitant infection of bacteria and amoeba in the dysentery cases [23]. However, in 1991 Lafleur described the presence of this amoeba in bacteriologically sterile liver abscess pus, thus signifying its pathogenic potential [24].

By 1903, the name '*Entamoeba histolytica'* was established for the species causing dysentery and a range of other diseases such as amebic liver abscess, cerebral and genitourinary amebiasis and respiratory tract infections [23]. By 1913, its life cycle and morphology were described [25]. But a clear picture of host parasite relation was yet to be established. Further, the experiments conducted by Walker and Sellards demonstrated that not all *E. histolytica* infections lead to invasive diseases and they confirmed the transmission through feco-oral route [25]. Different theories were proposed for this variable pathogenic outcome. Brumpt, in 1925, proposed the two species theory, describing that one remains commensal in the human gut, without producing any clinical symptoms and called it '*Entamoeba dispar,'* and the other one responsible for the invasive disease [26]. Although this hypothesis best described the epidemiology of *Entamoeba* but it was not widely accepted for next 50 years, till Sargeaunt et al. in 1982 confirmed it with isoenzyme analysis of 10,000 cases from different countries [27]. In 1991, Clark and Diamond also confirmed through study of ribosomal RNA genes that *E. histolytica* is actually two species, morphologically indistinguishable from each other, one of which causes disease [28]. Further through different immunological, biochemical and genetic studies *E. dispar* was formally accepted as a close but distinct species of *E. histolytica* [29-31].

It is estimated that approximately 12% of the world's population is infected with *E. histolytica* and *E. dispar* with the latter being significantly more common [11]. After the recognition of *E. dispar* as a different species, many studies were conducted comparing it with related species *E. histolytica* [32-36]. The first report of the isolation of *E. dispar* from the symptomatic cases came from Brazil and the isolated strain was able to produce lesions similar to *E. histolytica* in animal models [37-38]. Additionally, *E. dispar* has been found to be associated with dysenteric colitis as well as non-dysenteric colitis cases [37-39]. Also, the presence of *E. dispar* DNA sequences in patients with amebic liver abscess suggests that it may be involved in the lesion formation in human liver and intestine [40].

In 1941, Tshalaia isolated an *Entamoeba* species similar to *E. histolytica* and *E. dispar* from the Moscow sewage water and it was named as '*Entamoeba moshkovskii'* [41]. This species was then reported from many different countries [42, 43] and was considered as a free-living commensal. Two decades later, *E. moshkovskii* was reported from a patient presenting with diarrhea, epigastric pain and weight loss [44] and the pathogenic potential of this species came into light. After this, *E. moshkovskii* has been reported from human specimen from different countries around the world, including South Africa, Italy, North America, India, Iraq, Bangladesh [45-48]. A few studies have also shown the correlation between the clinical illness and presence of *E. moshkovskii* as pathogen. A report in 2012 demonstrated *E. moshkovskii* as the causative agent of the diarrhea in the cases of infants highlighting the need for further investigation of the pathogenic potential of this species [5].

Recently the analysis of the fecal samples of children positive for the *Entamoeba* infections through conventional methods but negative for any pathogenic *Entamoeba* species through molecular technique like PCR led to the discovery of another *Entamoeba* species known as *Entamoeba bangladeshi* nov. sp., the name denoting the geographical origin of the first patient [49]. This species is morphologically identical to other members of this genus (*E. histolytica/dispar/moshkovskii*) and techniques such as light microscopy, conventional culture or transmission electron microscopy has not been able to reveal any additional feature which can help in the differentiation of this species [50]. Through sequence analysis it was found that *E. bangladeshi* was more related to non-pathogenic *E. dispar*.

However, it was more closely related to *E. histolytica* than *E. moshkovskii* [49]. *E. bangladeshi* has the ability to grow at 37^{0}C and 25^{0}C like *E. moshkovskii* and this feature distinguishes it from *E. histolytica* and *E. dispar*. *E. bangladeshi* has been found to be associated with asymptomatic as well as from the diarrheal cases [49]. Recently an African study for the first time reported the presence of *E. bangladeshi* in samples collected outside Bangladesh [51].

EPIDEMIOLOGY

Entamoeba is much more commonly found in the developing countries as compared to the industrialized nations [52]. Most common *Entamoeba* infections occur in Asia, Africa, Central and South America [53]. In developing nations, the spread of this infection is due to fecal oral spread via contaminated water and food [10]. However, in developed nations immigrants, travelers and men who have sex with men (MSM) are found to be at greater risk of *Entamoeba* infections [54]. A report based on studies conducted in 47 countries showed the overall molecular prevalence of *Entamoeba* species to be 3.55% in humans worldwide, ranging from 1.72% in Oceania to 21.58% in North America [53]. However, more molecular studies are required to clearly understand the *Entamoeba* prevalence structure and distribution globally.

Developed World

The *Entamoeba* infections in the United States have been mostly associated with the returning travelers or the immigrants from the endemic regions [55]. Other industrialized nations such as part of Europe, North America, Asia and Australia have showed bisexuals and men who have sex with men (MSM) to be at higher risk of *Entamoeba* infections as compared to the general population [56-58]. Taiwan and Japan have reported HIV infected MSM to be prone for the invasive *Entamoeba* infections [57, 59-

60]. In Italy, the prevalence of *Entamoeba* spp. among the HIV population was screened and it was found that none of the isoenzyme pattern were positive for pathogenic strains [61]. A cohort study conducted in Germany among the returning travelers with gastrointestinal discomfort showed the prevalence of *E. histolytica* and *E. dispar* in 9.7% and 88.3% respectively [62]. In Canada, a small-scale study conducted on 66 patients positive for *Entamoeba* species showed only 2 (0.03%) to be positive for *E. histolytica* [63] It is important to note that in developed nations the incidence of *Entamoeba* species can be high but only a minor proportion of the virulent *E. histolytica* is reported. Similar results have been seen in other developed nations such as Australia (0.04%), Germany (0.1%) and Sweden (0.06%) [64-65].

Developing Countries

In developing countries, the burden of *Entamoeba* infections is difficult to quantify as it can be affected by a number of factors such as geographical region, study design, sample size, sensitivity and specificity of the diagnostic approach used [17]. In Malaysia, a study conducted among the Orang Asli settlement showed the prevalence of *E. histolytica* and *E. dispar* to be 13.2% and 9.9% respectively using the nested PCR and real time PCR technique [66]. A study conducted in Yemen among the rural communities showed the presence of *Entamoeba* species in 53.6% of the cases through microscopy, among which the prevalence of *E. histolytica* and *E. dispar* was found to be 20.2% and 15.7% respectively [67]. In India, the prevalence of *E. histolytica* was found to be 13.7% by PCR in a cross-sectional survey among the Northeastern states [68]. Similarly, the prevalence of *E. histolytica* has been reported to be 11% in the 7 provinces of China [69].

In Bangladesh diarrheal diseases are a leading cause of death in children [3]. Studies report that by the age of 5 years, 50% of the children in Bangladesh have the serological evidence for *Entamoeba* infections [3]. The prevalence of the different *Entamoeba* species in asymptomatic school children in Columbia was found to be 23.2% and 0.55% for *E. dispar* and *E. histolytica* respectively [70]. High prevalence of *Entamoeba* infections (52.8%) among the school children (age: 7-15 years) has also been reported from the Indonesia [71]. These studies point towards the higher risk of *Entamoeba* infections in school children in the developing countries than the general population. A large Global Enteric Multi-Center Study (GEMS) conducted across the sub-Saharan Africa and South Asia reported *E. histolytica* to be 1 out of top 10 cause of moderate to severe diarrhea among the children of five years of age [72].

Among the cases of gastrointestinal (GI) discomfort patients attending hospitals, a prevalence rate of 7% to 12.2% have been detected for *Entamoeba* infections through different diagnostic methods in India [73-75]. A study from Iran showed lower (1.6%) prevalence rate of *Entamoeba* species in the patients with GI discomfort, though majority of them were positive for *E. histolytica* [76]. Contrary to this, another study from the same country reported 24.4% prevalence of *E. histolytica/dispar* complex among the patients with GI discomfort [77]. A report from Nigeria conducted on GI discomfort patients showed a prevalence rate of 24.4% for *E. histolytica/dispar* complex [78]. A study from Pakistan has reported the prevalence rate of 23.1% of *E. histolytica* among the same population [79]. The other studies from developing world showing the prevalence of *Entamoeba* species has been listed in Table 1.

A number of studies have reported the association of the *E. moshkovskii* with the clinical symptoms. The first evidence of the *E. moshkovskii* in India was reported in Puducherry from the patients attending the hospital and a prevalence rate of 2.2% was detected [45]. Another study from Kenya reported the presence of *E. moshkovskii* in symptomatic as well as asymptomatic cases with a prevalence of 19.5% [96]. Recently a study from West Iran reported the presence of *E. moshkovskii* in the GI discomfort patients with a prevalence rate of 0.22% [97]. From Malaysia the first evidence of *E. moshkovskii* came in 2012 among the community with a 12.3% prevalence rate [98].

Table 1. Recent studies from the developing countries showing the prevalence of *Entamoeba* species in different study population

Country	Sample	Study population	Prevalence	*Entamoeba* spp.	Method of detection	Reference
Iraq	Stool	Children (1 month – 12 years)	9.80%	*E. histolytica*	Microscopy	[80]
Malaysia	Stool	Community population	20.4%	*E. histolytica*	PCR	[81]
			26.5%	*E. dispar*		
			20.4%	*E. moshkovskii*		
Nepal	Stool	Community population	0.7%	*E. histolytica*	PCR	[82]
			5.6%	*E. dispar*		
			0%	*E. moshkovskii*		
Qatar	Pus aspirate	Liver abscess patients	16.4%	*E. histolytica*	Indirect Hemagglutination test	[83]
Iran	Stool	Mentally retarded	15%	*E. histolytica/ dispar*	Microscopy	[84]
South Africa	Stool	GI discomfort patients	4.1%	*E. histolytica*	Multiplex PCR	[85]
			14.7%	*E. dispar*		
			15.9%	*E. moshkovskii*		
Malaysia	Pus aspirate	Liver abscess patients	76.7%	*E. histolytica*	Real time PCR	[86]

Table 1. (Continued)

Country	Sample	Study population	Prevalence	*Entamoeba* spp.	Method of detection	Reference
Nigeria	Stool	Children (1 month–14 years)	16%	*E. histolytica*	Microscopy	[87]
Country	Sample	Study population	Prevalence	*Entamoeba* spp.	Method of detection	Reference
Bangladesh	Stool	Diabetic patients	15.9%	*E. histolytica*	ELISA	[88]
South Africa	Stool	HIV positive patients	34%	*E. histolytica*	ELISA	[89]
Palestine	Stool	University female students	7.5%	*E. histolytica/ dispar*	Microscopy	[90]
Mexico	Stool	School children	1.8%	*E. histolytica*	Real time PCR	[91]
Ecuador	Stool	GI discomfort patients	2.8%	*E. histolytica*	Real time PCR	[92]
			69.8%	*E. dispar*		
Ethiopia	Stool	Community	1.7%	*E. histolytica*	Nested multiplex PCR	[93]
			42.2%	*E. dispar*		
Nigeria	Stool	Pregnant women	0.8%	*E. histolytica*	Microscopy	[94]
Tanzania	Stool	Community population	11.8%	*E. histolytica*	PCR	[95]

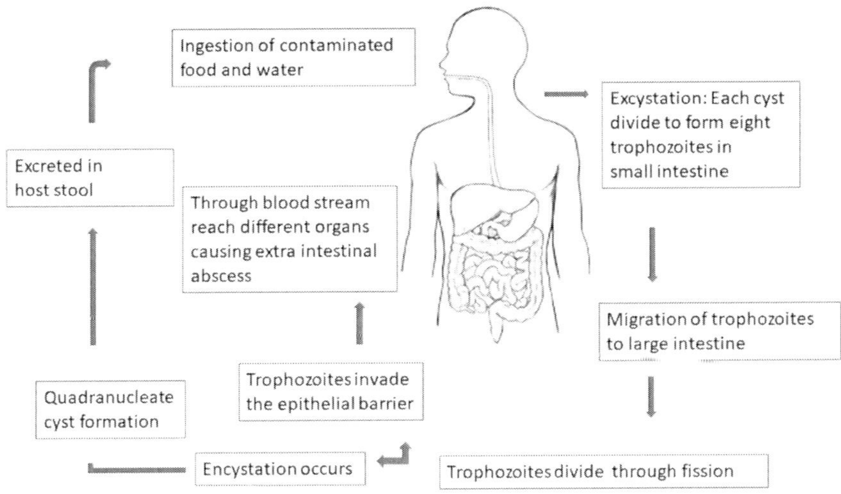

Figure 1. Life cycle of *E. histolytica* in human host.

The prevalence of *E. bangladeshi* has been restricted till date from stool samples of children of Bangladesh only, however, recently a study reported prevalence of *E. bangladeshi* (4.75%) in the stool samples of the patients attending gastroenterology clinics in South Africa [51]. There is no other report of incidence of *E. bangladeshi* from any other geographical region.

PATHOGENICITY

The incubation period in general for *E. histolytica* infections is 1 to 4 weeks [99]. However, it can vary from months to years [100]. The wide spectrum of infection ranges from asymptomatic colonization to fulminant colitis to extraintestinal abscess [101]. Figure 1 shows the life cycle of *E. histolytica* in human host. Majority of *E. histolytica* infections remain asymptomatic and merely 10-20% results in symptomatic cases [102-103]. The reason for such a partiality in virulence remains poorly understood. However, it is hypothesized to be the interplay of different factors related to parasite, host and environment [17]. Associated microflora has also been known to contribute to the differential pathogenesis of this parasite. Recently, it was found that *Prevotella* enhances the virulence of *E. histolytica* in the diarrheal cases [51, 104-106] and in our setting also we found, the significant association between presence of *Prevotella* and recurrence of amebic liver abscess [our unpublished data]. The risk factors associated with *E. histolytica* infection severity includes young age, alcoholism, malnutrition and corticosteroid use [107-108].

Luminal Amoebiasis

In upto 80 - 90% of the *E. histolytica* infections there are no clinical symptoms [100, 102-103]. The parasite may colonize the gut of the host for months and years without any external symptom of disease. Individuals with *E. histolytica* infections develop serum antibodies even in the asymptomatic cases in the absence of invasive disease. However, in the asymptomatic *E. dispar* infections, no anti-amebic antibodies are found [109-110]. Screening of asymptomatic population should be considered in those who have a travel history to endemic regions like Indian subcontinent, tropical and central region of Africa and South America [51]. Individuals with asymptomatic infection should be treated with a luminal agent to prevent any further transmission of the disease [111].

Amebic Colitis

Amoebic colitis has a range of symptoms from mild diarrhoea to severe dysentery. A characteristic feature of amoebic colitis is watery or bloody diarrhoea along with abdominal pain, tenderness and weight loss [51]. It is tough to distinguish amoebic colitis from inflammatory bowel disease even by using endoscopy, imaging or markers [112]. Presence of Charcot-Leyden crystals, absence of leukocytes and occult blood in stool are the main finding of the acute stage of the disease [113]. The best diagnostic technique includes antigen detection and molecular detection through polymerase chain reaction [13, 35, 113].

One of the life-threatening manifestations of amoebic colitis is fulminant infection which results in perforation, peritonitis, necrosis and toxic megacolon [114-116]. The mortality rate of fulminant colitis is 40% which may increase up to 89% if concomitant liver abscess is present [117-118]. Use of corticosteroid, alcoholism, malignancy, pregnancy has been described as a factor responsible for fulminant colitis [112]. Such cases are unresponsive to anti-amoebic treatment and need urgent surgical intervention [112, 119].

Further, ameboma is another complication associated with amoebic colitis. Usually confused with colonic cancer, this tumour-like granulation tissue tends to present with pain and swelling and symptoms of bowel obstruction [17, 120-121]. Ameboma is generally result of untreated or partially treated amoebic colitis [121]. Cutaneous amoebiasis [122-123] and rectovaginal fistulas [124] are other complications associated with intestinal amoebiasis.

Extra-Intestinal Amoebiasis

Amebic liver abscess (ALA) is the most common extra intestinal manifestation of *E. histolytica* [108, 125-126]. The most common symptoms associated with this disease includes fever, abdominal pain in right upper quadrant, tenderness, weight loss [127]. Diarrhoea has also been seen in 40% of the cases, however, the presence of the parasite in the stool samples of ALA patients is often not detectable [10]. ALA shows sex discrimination and more commonly found in adult males [126,128]. Leucocytosis, anaemia and elevated level of alkaline phosphatase are most commonly associated with ALA [129-130]. Solitary abscess is commonly found; however, multiple abscess has also been reported [125, 131]. The presentation of ALA as 'thick anchovy sauce' has been quite overstressed in past as other colours such as brown, dirty yellow or ivory have also been seen associated with ALA [126]. Although, the governing factors leading to invasive form of amoebiasis is unknown it has been suggested that production of different profiles of cytokines, reactive oxygen species (ROS) and nitric oxide provides the suitable environment for the formation of ALA [132].

The rupture of abscess into the pericardium is a complication associated with ALA with high mortality [133-135]. Symptoms of ruptured ALA include edema, severe chest pain and shortness of breath [135]. Bronchohepatic fistula, empyema and inferior vena cava (IVC) thrombosis are other complications associated with ALA [118, 136]. Rare cases of liver abscess are accompanied by brain abscess (<0.1%) which shows rapid progressive disease finally resulting in death [137].

Pulmonary abscess can occur as an extension of ALA or it can also spread from intestinal lesions by blood [138-139]. Lungs are the second most common site of extra intestinal infection of *E. histolytica* after liver [140-142]. Symptoms of amebic lung abscess include fever, hemoptysis upper quadrant pain [141].

Diagnostic Approaches

The epidemiology of the *Entamoeba* species remains uncertain as majority of the data have been obtained through methods which are incapable of distinguishing the morphologically identical species. In developing world, *Entamoeba* infections are a major cause of morbidity and mortality [143]. Even in developed countries travelers, immigrants and MSM are at high risk [55-60, 144]. Thus, early and reliable detection of the *Entamoeba* infection will improve the public health issues and accelerate the control measures. Laboratory diagnosis of the *Entamoeba* infections are based on parasitological examination, immunological techniques and molecular methods.

Due to the presence of morphological identical species of *Entamoeba* genus, the World Health Organisation (WHO) has recommended the application of advanced techniques for specific diagnosis [11].

Microscopy

The visual demonstration of the cysts/trophozoites of the *Entamoeba* species in clinical samples is performed by microscopic examination. In spite of the fact that it cannot differentiate between the morphological identical species of *Entamoeba* genus including *E. histolytica*, *E. dispar, E. moshkovskii* and *E. bangladeshi*, microscopy is the most frequently used diagnostic technique especially in developing countries [17]. Microscopic technique in diagnosis of *Entamoeba* species includes wet mount preparation, concentration technique and permanent staining for the identification in the clinical specimen. Wet saline mount has been found to be very insensitive (<10%) for the detection of *Entamoeba* species [145]. However, if the fresh samples are examined within an hour, motile trophozoites with or without ingested RBCs may be seen. Figure 2 shows the microscopic view of *Entamoeba* spp. from the fecal sample. Asymptomatic population carries only cysts in the fecal samples which can be demonstrated by using concentration techniques along with the permanent staining techniques such as methylene blue, Giemsa staining, trichome and iron hematoxylin. However, the use of iron hematoxylin staining has been limited due to use of mercuric chloride in its staining procedure as per the environmental considerations. Thus recently, a quick Kohn's one step staining method using BD CytoRichTM Blue for the pre-treatment of the samples has been reported [146].

Entamoeba trophozoites degenerate in minutes after exposure to air [147] so fixatives such as Poly-vinyl alcohol (PVA), Schaudinn's fluid, sodium acetate-acetic acid-formalin (SAF) is recommended to prevent the morphology of the cysts and trophozoites. Microscopy is the least reliable method for the diagnosis of *Entamoeba* species [148]. False positive results are very common due to misinterpretation of macrophages as trophozoites or presence of commensal species as pathogens. For improved diagnosis, microscopic examination of at least three stool samples over no more than ten days is recommended as the excretion of this parasite can be uneven and intermittent [148-149]. Microscopy is quite challenging for an inexperienced technician. Thus, the overall sensitivity and specificity of the microscopy in diagnosis of *Entamoeba* infections is quite low.

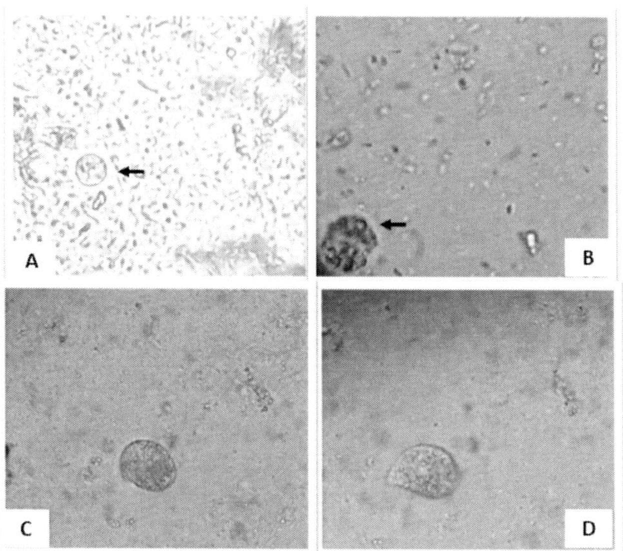

A: Cyst of *Entamoeba* spp. in stool; B: Trophozoite of *E. histolytica* in pus aspirate; C: Trophozoite of *E. histolytica* in stool; D: Trophozoite of *E. histolytica* with pseudopodia. (40X).

Figure 2. Microscopic view of cyst and trophozoite of *Entamoeba spp.* in fecal and liver aspirate samples.

Culture

Culture of the *Entamoeba* species can be done from fecal samples, biopsy specimen or liver abscess aspirates. However, the success rate for *Entamoeba* culture is about 60-70% only [150]. Culture media for the isolation of the *Entamoeba* species includes xenic and axenic systems. In xenic system, the parasite is grown in the presence of an undefined flora. It was first developed by Boeck and Drbohlav over 95 years ago in a diphasic egg slant. Of the variety of xenic media developed diphasic Locke egg media, TYSGM-9 and Robinson media are still commonly in use [150].

The cultivation of the *E. histolytica* through axenic system was first reported by Diamond in 1961 [151]. In axenic system the parasite is cultivated in the absence of any additional metabolizing cells. Presently, TYI-S-33 and YI-S are the most commonly used axenic media for the cultivation of the *E. histolytica* [152-153].

The chances of false negative results for *Entamoeba* culture are very high and it needs expertise, therefore, currently *Entamoeba* culture is not recommended for diagnosis and used only for research purpose.

E. dispar can be best grown in xenic media, however, very limited growth of *E. dispar* has been reported in monoxenic or axenic culture media suggesting that unlike *E. histolytica*, its close species *E. dispar* is not able to get nutrient from particle free media [154-155].

For the growth of the *E. moshlovskii* different media such as TYI-S-33 and TYSGM-9 has been used, however, additional 5-10% of the bovine serum provides better culture condition for this amoeba [152, 156].

The culture of parasites is not only expensive but also labor intensive. Additionally, it has sensitivity lower than the microscopy. Therefore, the use of culture is not recommended for the diagnosis purposes and now is only limited to research work.

Isoenzyme Analysis

Through isoenzyme analysis of cultured amoeba initially the differentiation of different *Entamoeba* species was possible [27]. Zymodeme consists of electrophoretic pattern of different enzymes such as hexokinase, malic enzyme, glucose phosphate isomerase and phosphoglucomutase enzyme [157]. Based on the electrophoretic pattern and mobilities of the different strains the zymodeme analysis is carried. There are 24 different established zymodeme patterns out of which 21 belongs to human isolates and 3 are from the experimentally cultured strains. *E. histolytica* and *E. dispar* can be distinguished based on isozyme analysis as they both genetically differ in hexokinase enzyme [158].

Before the development of the DNA based techniques, isoenzyme analysis was considered as the gold standard for the *Entamoeba* diagnosis. However, it has a number of associated disadvantages, as it needs expertise, it is labor and time consuming. Additionally, it also requires establishment of parasite culture which is a tedious process and over growth of any other fungi, protozoan or bacteria in culture may lead to uncertain results. Due to these reasons and the complexity of this method, it is not used in the routine diagnosis and remains confined only for research activities.

Antigen Detection Test

Different antigen-based enzyme linked immunosorbent assays (ELISAs) have been developed for the detection of the *E. histolytica* antigen in the fecal samples. Antigen based kits utilizes monoclonal antibodies against serine rich antigen of *E. histolytica* (Optimum S kit; Merlin Diagnostika, Bornheim-Hersel, Germany), Gal/GalNAc specific lectin of *E. histolytica* (*E. histolytica* II; TechLab, Blacksburg, VA and CELISA PATH kit, Cellabs, Brookvale, Australia) and a specific antigen (EHSA) from *E. histolytica/E. dispar* (ProSpecT ELISA (Remel Inc., Lenexa, KS, USA). There are also research based detection tests using monoclonal antibodies against lipophosphoglycan, lectin rich surafec antigen, pyruvate phosphate dikinase and an uncharacterized antigen of *E. histolytica* [159-161]. The most frequently used antigen detection test for *E. histolytica* is TechLab kit. The first generation of the kit was produced in 1993 which was highly conserved and specifically detects the Gal/GalNAc lectin of *E. histolytica*. The sensitivity and specificity of the test was found to be higher as compared to the microscopy and culture [148, 162]. However, later a few studies reported reduced sensitivity of the test for the detection and discrimination of *E. histolytica* [63, 103, 160]. Due to some limitation, the second generation of the kit was launched as TechLab *E. histolytica* II which has been used in many epidemiological studies. The salivary antigen has also been found to be a predictor of the

Entamoeba infections as a study showed presence of lectin in saliva has high specificity (97.4%) in early amebic colitis [163].

These ELISAs have been used in different studies for the detection of the *Entamoeba* species using clinical specimen such as faecal samples, serum and liver pus aspirates [13, 16, 32, 148, 161]. The antigen-based detection method is rapid and technically simple than other diagnostic approaches, however, the sensitivity and specificity of this method has given contradictory results in many epidemiological studies. To establish the actual worldwide distribution of the *Entamoeba* species molecular assay are better suited. However, in resource limited settings where DNA based molecular tests are not available, antigen-based ELISA can be a good option.

Antibody Detection Test

The antibody detection assay can be useful in the regions where *Entamoeba* infections are not common as it cannot differentiate between the past from current infection [110, 164]. Serological testing has been found to be very useful in the case of amebic liver abscess patients as in these patients, detectable number of parasites are not excreted in the stool samples [10]. Different serological assays have been developed for the detection of antibodies including immunohemagglutination (IHA), latex agglutination, amebic gel diffusion test, ELISA, counter-immuno-electrophoresis [10]. It has been found that the sensitivity of the antibody detection in serum of liver abscess patients is 100%, however, the immune response against *E. histolytica* infections have been found in asymptomatic population also. Serum IgG antibodies can persist for a long time even after the infection but IgM antibodies are short lived and can be detected in cases of current infection only.

Conventional Polymerase Chain Reaction (PCR)

DNA based approaches are a method of choice for epidemiological studies in developed world [165-167]. Use of PCR for the detection of *Entamoeba* species has been strongly recommended by WHO [11]. PCR targeting the small subunit rRNA gene has been used massively due to its presence in multiple copies of the extrachromosomal plasmid and it is reported to be more sensitive than any other diagnostic approach [160]. A wide variety of the PCR methods targeting different genes have been used to differentiate between the *Entamoeba* species. DNA extracted from either laboratory grown amebic culture or directly from the clinical samples can be used for the diagnosis of *Entamoeba* infection [168]. The first successful use of PCR for studying the epidemiology of the *Entamoeba* infection was carried out by Acuna-Soto et al. targeting the extra chromosomal circular DNA [32]. Primers targeting 125-KDa surface antigen, 29-KDa/30KDa antigen gene, SREPH gene, chitinase gene, cysteine proteinase gene, actin gene and haemolysin gene (HLY6) have been used in a number of studies [14, 169-173]. PCR-RFLP has been used in several studies as an approach to differentiate between the *Entamoeba* species. The first study for the discrimination of *E. histolytica* and *E. dispar* from a single gene amplicon through RFLP was reported in 1991 [172].

To increase the sensitivity of the PCR assay, a nested multiplex PCR has been developed for simultaneous detection and differentiation of the *Entamoeba* species in the clinical samples. The two nested protocol assays targeting the 16S like rRNA gene have been given by Jawaharlal Institute of Postgraduate Medical Education & Research, Puducherry, India [174] and International Centre for Diarrheal Diseases and Research, Dhaka, Bangladesh [46]. They have reported a diagnostic specificity of 100% for the detection of *Entamoeba* and simultaneous differentiation of the pathogenic species [46]. However, due to difficulty in DNA extraction from faecal samples, its cost and requirement of technical expertise, the application of PCR in routine diagnosis is limited.

Real Time PCR

Real time PCR is a quantitative method which allows the determination of the number of parasites and hence the microbial load in a clinical sample. It has enhanced sensitivity as compared to the conventional PCR with a capability to detect 0.1 cell per gram of stool [175]. Additionally, real time PCR has reduced risk of the amplicon contamination from the laboratory [176]. Real time PCR also allows the specific detection of amplicons by binding to labelled probes during the assay and thus enabling continuous monitoring of amplicons. The sequence used in the majority of the real time studies include rRNA gene as target.

Different studies have shown the use of real time PCR in the detection of the *E. histolytica* and *E. dispar* from the stool samples [177-179]. Qvarnstrom et al. used different probes for the detection of the *Entamoeba* infection and suggested that TaqMan method is more specific than SYBR Green approach [180]. The development of the multiplex real time PCR made it possible to identify, genotype and quantify the different pathogens in a single reaction [181]. A few multiplex PCR panels have been certified for the diagnosis of different pathogens in clinical samples [178].

Real time PCR is highly sensitive and specific but due to its high cost as compared to other approaches such as antigen-based test, microscopy and serological testing, developing regions with endemic *Entamoeba* infections are less likely to be benefitted from this advance technique.

Loop Mediated Isothermal Amplification Assay (LAMP)

The isothermal amplification is a rapid, sensitive and simple diagnostic method especially for low resources areas. Studies have shown the superior sensitivity of the LAMP over the DNA based nested PCR technique [182]. However, in a study the detection limit of the LAMP was found to be 5 parasites per reaction corresponding to 15.8 ng/μl DNA and that for conventional PCR is just 2 ng/μl DNA [183-184]. Later, a modification of the thermal assay was done using 18S rRNA gene with extra reaction accelerating primers and was called stem loop. A study showed that stem LAMP is better in detection of *Entamoeba* species as compared to the standard LAMP and conventional PCR [185].

Microarray

Microarray is sensitive and rapid technique for the detection of pathogens. The major steps involved in microarray includes, DNA extraction, amplification of DNA, hybridization of the labelled DNA with the probes on microarray and data analysis. Wang et al. had developed a microarray for the simultaneous detection of *E. histolytica* and *E. dispar* [186]. A study revealed that microarray could be used to differentiate *E. histolytica* from its closely related *E. dispar* along with the identification of the genes related to virulence for the genotypic and phenotypic related study [187]. Presently microarray is just a research tool and due to its high cost and is not used as a diagnostic approach.

Treatment

WHO recommends that all patients infected with *E. histolytica* should be treated [188]. Those with clinical symptoms should be administered an amoebicidal tissue active agent and a luminal cysticidal agent and in the case of asymptomatic cyst passers only latter should be given. The most common treatment of choice is nitroimidazoles which are recommended for amebic colitis as well as amebic liver abscess [17]. The most commonly used agent from this class is metronidazole. Metronidazole have been found to have action against the intestinal as well as extraintestinal infections, however, due to its high tissue penetration, it is more effective in extraintestinal cases [189]. Additionally, it was found that metronidazole can be mutagenic in bacteria and carcinogenic in mammal model over long period of time [190-191]. Other agents of nitroimidazole group includes tinidazole and ornidazole which have longer half-lives and shorter period of treatment. They tend to be better tolerated but the availability of these nitroimidazoles is limited [136]. A thiazolide agent, nitazoxanide, has also been used in few cases, to treat symptomatic cases of amoebiasis based on a trial from Egypt, however, a large population study is needed for validation of efficacy of this drug before using it as a treatment choice [192]. Emetine and hydroemetine have been found to be effective against the invasive stage of amoeba but presently their use has been discontinued owing to the high toxicity of the drug [189]. Emetine shows high efficiency when administered intramuscular but the use of this drug has been limited to those cases where the commonly used nitroimidazoles are not effective. Another drug, chloroquine has been found to be effective against the hepatic *Entamoeba* infections but not in the case of dysentery because of its higher absorption in the small intestine [193]. For treatment of luminal carriage, drugs with cysticidal activity include iodoquinol, paromomycin and diloxanide furoate. However, iodoquinol and diloxanide furoate has availability issues and paramomycin cannot be administered along with metronidazole as it causes diarrhea [194]. The recommended dosage of the drugs used along with duration and side effects has been given in Table 2.

As an advancement in this direction, Auranofin has emerged as an antiparasitic drug through automated high-throughput drug screening, which was formerly used to treat selected cases of arthritis [195]. But issues such as its effectiveness and safety concerns need to be resolved before its use as a drug for amoebiasis. A few bioactive compounds (curcumin and flavonoids) have shown potential as anti *E. histolytica* agent, however, there is a long way of their characterization and toxicity assessment before their clinical application [196]. In another approach, by using target-based drug screen, five compounds (rutilantin, rifabutin, pararosaniline pamoate, gentian violet and cetylpyridinium chloride) as *Entamoeba* heat shock protein 90 inhibitors have shown *E. histolytica* inhibition in micromolar range [197].

In case of amebic liver abscess along with drug therapy, ultrasound guided drainage is also recommended, but only if the size of the abscess is greater than 5cm [198]. It has been shown that in case of large abscesses needle aspiration along with metronidazole hastens the clinical improvement [199].

From last 60 years, no drug development has happened in the case of amoebiasis, despite of urgencies set by NIAID for drug development for treatment of category B pathogens. Thus, there is need for development of specific, effective and novel drug to combat *E. histolytica* infections.

Table 2. Recommended antiparasitic drugs in *Entamoeba* infections

Disease	Drug therapy	Dose	Duration	Side effects
Amebic colitis	Metronidazole (5-Nitroimidazole) followed by luminal agent*	750 mg for three times a day	7-10 days	Nausea, vomiting, diarrhea, abdominal discomfort, anorexia
Amebic liver abscess	Tinidazole (5-Nitroimidazole) followed by a luminal agent	2 g once daily	3-5 days	Primarily gastrointestinal symptoms and intolerance reaction with alcohol
	Metronidazole (5-Nitroimidazole) followed by luminal agent*	750 mg for three times a day	7-10 days	Nausea, vomiting, diarrhea, abdominal discomfort, anorexia
Asymptomatic intestinal colonization	*Paromycin (aminoglycoside)	25-30 mg/ kg three times a day	7 days	Diarrhea, gastrointestinal upset
	*Diloxanide furoate	500 mg three times a day	10 days	Nausea, vomiting, flatulence, pruritus, urticaris

PREVENTION

There is no vaccination for amoebiasis. Although Gal-lectin based vaccination appears to be promising in animal models and a review also suggests that other virulence factors such as serine rich *E. histolytica* protein has strong candidature as a vaccine [200]. A few other potential vaccine candidates include the pre-forming peptide, the Cysproteinase, the lectin and a Ser-rich plasma membrane protein but only the lectin and Ser-rich protein have been shown to immunize experimental animals against hepatic challenge with *E. histolytica* [201]. But there is a long way before it fits for human use. Without vaccines, the main control relies on prevention of transmission of amoebiasis through water and food safety, hand hygiene and avoiding feco-oral exposure. Before travel to endemic regions, individuals should be advised on proper food safety. Travel history is important in case of amoebiasis as amebic liver abscess and colitis can occur even after a decade of the travel to endemic countries. Along with this, households should be screened for prevention of the secondary infections as they are an important source of transmission of this disease.

FUTURE CHALLENGES

Amoebiasis remains a public health problem especially in impoverished population globally. The overall improvement in sanitation and socio-economic conditions are required long term goals. However, in a world where 2.3 billion populations do not have basic sanitation facilities and 1.8 billion people use fecal contaminated drinking water facilities, it is quite a challenging task. In the meantime, at least small interventions, for example, hand washing, better sanitary conditions, fly control should be examined in follow up to improve the conditions in endemic regions. Despite several constraints, much can be done with the available tool and knowledge for betterment of the current situation. Improved understanding will not only open new avenues for control of *E. histolytica* infections but will also reveal some novel way for vaccine development and prevention strategies.

References

[1] Baldauf, Sandra L. "An overview of the phylogeny and diversity of eukaryotes." *Journal of Systematics and Evolution* 46, no. 3 (2008): 263-273.

[2] Costa, Juliana de Oliveira, José Adão Resende, Frederico Ferreira Gil, Joseph Fabiano Guimarães Santos, and Maria Aparecida Gomes. "Prevalence of *Entamoeba histolytica* and other enteral parasitic diseases in the metropolitan region of Belo Horizonte, Brazil. A cross-sectional study." *Sao Paulo Medical Journal* 136, no. 4 (2018): 319-323.

[3] Haque, Rashidul, Ibnekarim M. Ali, and William A. Petri Jr. "Prevalence and immune response to *Entamoeba histolytica* infection in preschool children in Bangladesh." *The American journal of tropical medicine and hygiene* 60, no. 6 (1999): 1031-1034.

[4] Turkeltaub, Joshua A., Thomas R. McCarty III, and Peter J. Hotez. "The intestinal protozoa: emerging impact on global health and development." *Current opinion in gastroenterology* 31, no. 1 (2015): 38-44.

[5] Shimokawa, Chikako, Mamun Kabir, Mami Taniuchi, Dinesh Mondal, Seiki Kobayashi, Ibne Karim M. Ali, Shihab U. Sobuz et al. "*Entamoeba moshkovskii* is associated with diarrhea in infants and causes diarrhea and colitis in mice." *The Journal of infectious diseases* 206, no. 5 (2012): 744-751.

[6] Royer, Tricia L., Carol Gilchrist, Mamun Kabir, Tuhinur Arju, Katherine S. Ralston, Rashidul Haque, C. Graham Clark, and William A. Petri Jr. "*Entamoeba bangladeshi* nov. sp., Bangladesh." *Emerging infectious diseases* 18, no. 9 (2012): 1543.

[7] Oliveira, Fabrício Marcus Silva, Elisabeth Neumann, Maria Aparecida Gomes, and Marcelo Vidigal Caliari. "*Entamoeba dispar*: could it be pathogenic." *Tropical parasitology* 5, no. 1 (2015): 9.

[8] Shnawa, Bushra Hussain. "Molecular Diagnosis of *Entamoeba histolytica*, *Entamoeba dispar*, and *Entamoeba moshkovskii*: An

Update Review." *Annual Research & Review in Biology* (2017): 1-12.

[9] Burgess, Stacey L., and William A. Petri. "The intestinal bacterial microbiome and *E. histolytica* infection." *Current Tropical Medicine Reports* 3, no. 3 (2016): 71-74.

[10] Fotedar, Rashmi, Damien Stark, N. Beebe, D. Marriott, J. Ellis, and J. Harkness. "Laboratory diagnostic techniques for *Entamoeba* species." *Clinical microbiology reviews* 20, no. 3 (2007): 511-532.

[11] Amoebiasis, W. H. O. "Report on the WHO." *Pan American Health Organization/UNESCO Expert Consultation Mexico City Geneva-WHO. Wkly Epidemiol Rec* 72 (1997): 97-100.

[12] Pillai, Dylan R., Jay S. Keystone, Donald C. Sheppard, J. Dick MacLean, Douglas W. MacPherson, and Kevin C. Kain. "*Entamoeba histolytica* and *Entamoeba dispar*: epidemiology and comparison of diagnostic methods in a setting of nonendemicity." *Clinical Infectious Diseases* 29, no. 5 (1999): 1315-1318.

[13] Haque, Rashidul, Nasir Uddin Mollah, Ibne Karim M. Ali, Khorshed Alam, Aleida Eubanks, David Lyerly, and William A. Petri. "Diagnosis of amebic liver abscess and intestinal infection with the TechLab *Entamoeba histolytica* II antigen detection and antibody tests." *Journal of Clinical Microbiology* 38, no. 9 (2000): 3235-3239.

[14] Tachibana, H., S. Kobayashi, E. Okuzawa, and G. Masuda. "Detection of pathogenic *Entamoeba histolytica* DNA in liver abscess fluid by polymerase chain reaction." *International journal for parasitology* 22, no. 8 (1992): 1193-1196.

[15] Zaman, S., J. Khoo, S. W. Ng, R. Ahmed, M. A. Khan, R. Hussain, and V. Zaman. "Direct amplification of *Entamoeba histolytica* DNA from amoebic liver abscess pus using polymerase chain reaction." *Parasitology research* 86, no. 9 (2000): 724-728.

[16] Zengzhu, G., R. Bracha, Y. Nuchamowitz, W. Cheng, and D. Mirelman. "Analysis by Enzyme-Linked Immunosorbent Assay and PCR of Human Liver Abscess Aspirates from Patients in China for

Entamoeba histolytica." *Journal of clinical microbiology* 37, no. 9 (1999): 3034-3036.

[17] Shirley, Debbie-Ann T., Laura Farr, Koji Watanabe, and Shannon Moonah. "A review of the global burden, new diagnostics, and current therapeutics for amebiasis." In *Open forum infectious diseases*, vol. 5, no. 7, p. ofy161. US: Oxford University Press, 2018.

[18] Paulish-Miller, Teresa E., Peter Augostini, Jessica A. Schuyler, William L. Smith, Eli Mordechai, Martin E. Adelson, Scott E. Gygax, William E. Secor, and David W. Hilbert. "Trichomonas vaginalis metronidazole resistance is associated with single nucleotide polymorphisms in the nitroreductase genes ntr4Tv and ntr6Tv." *Antimicrobial agents and chemotherapy* 58, no. 5 (2014): 2938-2943.

[19] Creemers-Schild, D., P. J. J. van Genderen, L. G. Visser, J. J. van Hellemond, and P. J. Wismans. "Recurrent amebic liver abscesses over a 16-year period: a case report." *BMC research notes* 9, no. 1 (2016): 472.

[20] Ng, C. H., Lawrence Lai, K. S. Ng, and K. K. Li. "Relapse of amoebic infection 10 years after the infection." *Hong Kong Medical Journal* 17 (2011): 71-73.

[21] Singal, Dinesh Kumar, Amit Mittal, and Anukalp Prakash. "Recurrent amebic liver abscess." *Indian Journal of Gastroenterology* 31, no. 5 (2012): 271-273.

[22] Stilwell, George G. "Amebiasis: its early history." *Gastroenterology* 28, no. 4 (1955): 606-622.

[23] Kean, B. H. "A history of amebiasis." *Amebiasis: human infection by Entamoeba histolytica. John Wiley & Sons, Inc., New York, NY* (1988): 1-10.

[24] Jackson, T. F. F. G. "Epidemiology." *Tropical Medicine: Science and Practice Amebiasis (2000)*: 47-63.

[25] Walker, Ernest Linwood, and Andrew Watson Sellards. "Experimental entamoebic dysentery." *Philippine J. of Science. Sect. B. Trop. Med.* 7, no. 4 (1913).

[26] Brumpt, E. "Eude sommarire de l'" *Entamoeba dispar*" n. sp. Amibe a kystes quadrinuclees, parasite de l'homme." *Bull Acad Med* 94 (1925): 943-952. ["Summary study of the" *Entamoeba* dispar "n. sp. Quadrinucleic cyst amoeba, parasite of man."]

[27] Sargeaunt, P. G., J. E. Williams, and J. D. Grene. "The differentiation of invasive and non-invasive *Entamoeba histolytica* by isoenzyme electrophoresis." *Transactions of the Royal Society of Tropical Medicine and Hygiene* 72, no. 5 (1978): 519-521.

[28] Clark, C. Graham, and Louis S. Diamond. "Ribosomal RNA genes of 'pathogenic' and 'nonpathogenic' *Entamoeba histolytica* are distinct." *Molecular and biochemical parasitology* 49, no. 2 (1991): 297-302.

[29] Diamond, L. S., and Clark, C. Graham. "A Redescription of *Entamoeba histolytica* Schaudinn, 1903 (Emended Walker, 1911) Separating It From *Entamoeba dispar* Brumpt, 1925 1." *Journal of Eukaryotic Microbiology* 40, no. 3 (1993): 340-344.

[30] Sargeaunt, P. G., and J. E. Williams. "Electrophoretic isoenzyme patterns of the pathogenic and non-pathogenic intestinal amoebae of man." *Transactions of the Royal Society of Tropical Medicine and Hygiene* 73, no. 2 (1979): 225-227.

[31] Strachan, W. D., W. M. Spice, P. L. Chiodini, A. H. Moody, and J. P. Ackers. "Immunological differentiation of pathogenic and non-pathogenic isolates of *Entamoeba histolytica*." *The lancet* 331, no. 8585 (1988): 561-563.

[32] Acuna-Soto, Rodolfo, John Samuelson, Paola De Girolami, Lauren Zarate, Francisco Millan-Velasco, Gary Schoolnick, and Dyann Wirth. "Application of the polymerase chain reaction to the epidemiology of pathogenic and nonpathogenic *Entamoeba histolytica*." *The American journal of tropical medicine and hygiene* 48, no. 1 (1993): 58-70.

[33] Rivera, Windell L., Hiroshi Tachibana, Mary Rose Agnes Silva-Tahat, Haruki Uemura, and H. Kanbara. "Differentiation of *Entamoeba histolytica* and *E. dispar* DNA from cysts present in stool specimens by polymerase chain reaction: its field application

in the Philippines." *Parasitology research* 82, no. 7 (1996): 585-589.

[34] Britten, Dawn, Stuart M. Wilson, Ruth McNerney, Anthony H. Moody, Peter L. Chiodini, and John P. Ackers. "An improved colorimetric PCR-based method for detection and differentiation of *Entamoeba histolytica* and *Entamoeba dispar* in feces." *Journal of Clinical Microbiology* 35, no. 5 (1997): 1108-1111.

[35] Haque, Rashidul, I. K. M. Ali, S. Akther, and William A. Petri. "Comparison of PCR, isoenzyme analysis, and antigen detection for diagnosis of *Entamoeba histolytica* infection." *Journal of Clinical Microbiology* 36, no. 2 (1998): 449-452.

[36] Verweij, J. J., J. Blotkamp, E. A. T. Brienen, A. Aguirre, and A. M. Polderman. "Differentiation of *Entamoeba histolytica* and *Entamoeba dispar* cysts using polymerase chain reaction on DNA isolated from faeces with spin columns." *European Journal of Clinical Microbiology and Infectious Diseases* 19, no. 5 (2000): 358-361.

[37] Martinez, Ana Maria Barral de, Maria Aparecida Gomes, João da Costa Viana, Alvaro José Romanha, and Edward Felix Silva. "Isoenzyme profile as parameter to differentiate pathogenic strains of *Entamoeba histolytica* in Brazil." *Revista do Instituto de Medicina Tropical de São Paulo* 38, no. 6 (1996): 407-412.

[38] Gomes, M. A., E. F. Silva, A. M. Macedo, A. R. Vago, and M. N. Melo. "LSSP–PCR for characterization of strains of *Entamoeba histolytica* isolated in Brazil." *Parasitology* 114, no. 6 (1997): 517-520.

[39] Graffeo, Rosalia, Carola Maria Archibusacci, Silvia Soldini, Lucio Romano, and Luca Masucci. "*Entamoeba dispar*: a rare case of enteritis in a patient living in a nonendemic area." *Case Reports in Gastrointestinal Medicine* 2014 (2014).

[40] Ximénez, Cecilia, Rene Cerritos, Liliana Rojas, Silvio Dolabella, Patricia Morán, Mineko Shibayama, Enrique González et al. "Human amebiasis: breaking the paradigm?." *International Journal*

of Environmental Research and Public Health 7, no. 3 (2010): 1105-1120.

[41] Tshalaia, L. "A Species of *Entamoeba* detected in Sewage." *Meditsinskaya Parazitologiya i Parazitarnye Bolezni* 10, no. 2 (1941).

[42] Clark, C. Graham, and Louis S. Diamond. "The Laredo strain and other '*Entamoeba histolytica*-like' amoebae are *Entamoeba moshkovskii*." *Molecular and biochemical parasitology* 46, no. 1 (1991): 11-18.

[43] Scaglia, M., S. Gatti, M. Strosselli, V. Grazioli, and M. R. Villa. "*Entamoeba* moshkovskll (Tshalaia, 1941) Morpho-biological characterization of new strains isolated from the environment, and a review of the literature." *Annales de parasitologie humaine et comparée* 58, no. 5 (1983): 413-422.

[44] Dreyer, D. A. "Growth of a strain of *Entamoeba histolytica* at room temperature." *Texas Reports on Biology and Medicine* 19, no. 2 (1961): 393-6.

[45] Parija, Subhash Chandra, and Krishna Khairnar. "*Entamoeba moshkovskii* and *Entamoeba dispar*-associated infections in Pondicherry, India." *Journal of Health, Population and Nutrition* (2005): 292-295.

[46] Ali, Ibne Karin M., Mohammad Bakhtiar Hossain, Shantanu Roy, Patrick F. Ayeh-Kumi, William A. Petri Jr, Rashidul Haque, and C. Graham Clark. "*Entamoeba moshkovskii* infections in children in Bangladesh." *Emerging infectious diseases* 9, no. 5 (2003): 580.

[47] Fotedar, R., D. Stark, N. Beebe, D. Marriott, J. Ellis, and J. Harkness. "PCR detection of *Entamoeba histolytica*, *Entamoeba dispar*, and *Entamoeba moshkovskii* in stool samples from Sydney, Australia." *Journal of Clinical Microbiology* 45, no. 3 (2007): 1035-1037.

[48] Tanyuksel, Mehmet, Mustafa Ulukanligil, Zeynep Guclu, Engin Araz, Ozgur Koru, and William A. Petri Jr. "Two cases of rarely recognized infection with *Entamoeba moshkovskii*." *The American journal of tropical medicine and hygiene* 76, no. 4 (2007): 723-724.

[49] Royer, Tricia L., Carol Gilchrist, Mamun Kabir, Tuhinur Arju, Katherine S. Ralston, Rashidul Haque, C. Graham Clark, and William A. Petri Jr. "*Entamoeba bangladeshi* nov. sp., Bangladesh." *Emerging infectious diseases* 18, no. 9 (2012): 1543.

[50] Gilchrist, Carol A. "*Entamoeba bangladeshi*: An insight." *Tropical parasitology* 4, no. 2 (2014): 96.

[51] Ngobeni, Renay, Amidou Samie, Shannon Moonah, Koji Watanabe, William A. Petri Jr, and Carol Gilchrist. "*Entamoeba* species in South Africa: correlations with the host microbiome, parasite burdens, and first description of *Entamoeba bangladeshi* outside of Asia." *The Journal of infectious diseases* 216, no. 12 (2017): 1592-1600.

[52] Walsh, Julia A. "Problems in recognition and diagnosis of amebiasis: estimation of the global magnitude of morbidity and mortality." *Reviews of infectious diseases* 8, no. 2 (1986): 228-238.

[53] Cui, Zhaohui, Junqiang Li, Yuancai Chen, and Longxian Zhang. "Molecular epidemiology, evolution, and phylogeny of *Entamoeba* spp." *Infection, Genetics and Evolution* 75 (2019): 104018.

[54] Nozaki, Tomoyoshi, Severa RN Motta, Tsutomu Takeuchi, Seiki Kobayashi, and Peter G. Sargeaunt. "Pathogenic zymodemes of *Entamoeba histolytica* in Japanese male homosexual population." *Transactions of the Royal Society of Tropical Medicine and Hygiene* 83, no. 4 (1989): 525-525.

[55] Swaminathan, Ashwin, Joseph Torresi, Patricia Schlagenhauf, Karin Thursky, Annelies Wilder-Smith, Bradley A. Connor, Eli Schwartz, Jay Keystone, and Daniel P. O'Brien. "A global study of pathogens and host risk factors associated with infectious gastrointestinal disease in returned international travellers." *Journal of Infection* 59, no. 1 (2009): 19-27.

[56] James, Rodney, Joel Barratt, Deborah Marriott, John Harkness, and Damien Stark. "Seroprevalence of *Entamoeba histolytica* infection among men who have sex with men in Sydney, Australia." *The American journal of tropical medicine and hygiene* 83, no. 4 (2010): 914-916.

[57] Hung, Chien-Ching, Sui-Yuan Chang, and Dar-Der Ji. "*Entamoeba histolytica* infection in men who have sex with men." *The Lancet infectious diseases* 12, no. 9 (2012): 729-736.

[58] Stark, Damien J., Rashmi Fotedar, John T. Ellis, and John L. Harkness. "Locally acquired infection with *Entamoeba histolytica* in men who have sex with men in Australia." *Medical Journal of Australia* 185, no. 8 (2006): 417-417.

[59] Tsai, Jih-Jin, Hsin-Yun Sun, Liang-Yin Ke, Keh-Sung Tsai, Sui-Yuan Chang, Szu-Min Hsieh, Chin-Fu Hsiao, Jeng-Hsien Yen, Chien-Ching Hung, and Shan-Chwen Chang. "Higher seroprevalence of *Entamoeba histolytica* infection is associated with human immunodeficiency virus type 1 infection in Taiwan." *The American journal of tropical medicine and hygiene* 74, no. 6 (2006): 1016-1019.

[60] Hung, Chien-Ching, Dar-Der Ji, Hsin-Yun Sun, Ya-Tien Lee, Shui-Yuan Hsu, Sui-Yuan Chang, Cheng-Hsin Wu et al. "Increased risk for *Entamoeba histolytica* infection and invasive amebiasis in HIV seropositive men who have sex with men in Taiwan." *PLoS Negl Trop Dis* 2, no. 2 (2008): e175.

[61] Gatti, Simonetta, Claudia Cevini, Chiara Atzori, Simona Muratori, Roberto Zerboni, Mario Cusini, and Massimo Scaglia. "Non-pathogenic *Entamoeba histolytica* in Italian HIV-infected homosexuals." *Zentralblatt für Bakteriologie* 277, no. 3 (1992): 382-388.

[62] Herbinger, K-H., E. Fleischmann, C. Weber, P. Perona, T. Löscher, and G. Bretzel. "Epidemiological, clinical, and diagnostic data on intestinal infections with *Entamoeba histolytica* and *Entamoeba dispar* among returning travelers." *Infection* 39, no. 6 (2011): 527-535.

[63] Gonin, Patrick, and Louise Trudel. "Detection and differentiation of *Entamoeba histolytica* and *Entamoeba dispar* isolates in clinical samples by PCR and enzyme-linked immunosorbent assay." *Journal of Clinical Microbiology* 41, no. 1 (2003): 237-241.

[64] Herbinger, K-H., E. Fleischmann, C. Weber, P. Perona, T. Löscher, and G. Bretzel. "Epidemiological, clinical, and diagnostic data on intestinal infections with *Entamoeba histolytica* and *Entamoeba dispar* among returning travelers." *Infection* 39, no. 6 (2011): 527-535.

[65] Lebbad, Marianne, and Staffan G. Svärd. "PCR differentiation of *Entamoeba histolytica* and *Entamoeba dispar* from patients with amoeba infection initially diagnosed by microscopy." *Scandinavian journal of infectious diseases* 37, no. 9 (2005): 680-685.

[66] Lau, Yee Ling, Claudia Anthony, Siti Aminah Fakhrurrazi, Jamaiah Ibrahim, Init Ithoi, and Rohela Mahmud. "Real-time PCR assay in differentiating *Entamoeba histolytica*, *Entamoeba dispar*, and *Entamoeba moshkovskii* infections in Orang Asli settlements in Malaysia." *Parasites & vectors* 6, no. 1 (2013): 1-8.

[67] Al-Areeqi, Mona A., Hany Sady, Hesham M. Al-Mekhlafi, Tengku Shahrul Anuar, Abdulelah H. Al-Adhroey, Wahib M. Atroosh, Salwa Dawaki et al. "First molecular epidemiology of *Entamoeba histolytica*, *E. dispar* and *E. moshkovskii* infections in Yemen: different species-specific associated risk factors." *Tropical Medicine & International Health* 22, no. 4 (2017): 493-504.

[68] Nath, Joyobrato, Sankar Kumar Ghosh, Baby Singha, and Jaishree Paul. "Molecular epidemiology of amoebiasis: a cross-sectional study among North East Indian population." *PLoS neglected tropical diseases* 9, no. 12 (2015): e0004225.

[69] Yang, Bin, Yingdan Chen, Liang Wu, Longqi Xu, Hiroshi Tachibana, and Xunjia Cheng. "Seroprevalence of *Entamoeba histolytica* infection in China." *The American journal of tropical medicine and hygiene* 87, no. 1 (2012): 97-103.

[70] López, Myriam Consuelo, Cielo M. León, Jairo Fonseca, Patricia Reyes, Ligia Moncada, Mario J. Olivera, and Juan David Ramírez. "Molecular epidemiology of *Entamoeba*: first description of *Entamoeba moshkovskii* in a rural area from Central Colombia." *PLoS One* 10, no. 10 (2015): e0140302.

[71] Matsumura, Takahiro, Joko Hendarto, Tetsushi Mizuno, Din Syafruddin, Hisao Yoshikawa, Makoto Matsubayashi, Taro Nishimura, and Masaharu Tokoro. "Possible pathogenicity of commensal *Entamoeba hartmanni* revealed by molecular screening of healthy school children in Indonesia." *Tropical medicine and health* 47, no. 1 (2019): 7.

[72] Kotloff, Karen L., James P. Nataro, William C. Blackwelder, Dilruba Nasrin, Tamer H. Farag, Sandra Panchalingam, Yukun Wu et al. "Burden and aetiology of diarrhoeal disease in infants and young children in developing countries (the Global Enteric Multicenter Study, GEMS): a prospective, case-control study." *The Lancet* 382, no. 9888 (2013): 209-222.

[73] Verma, Anil Kumar, Ravi Verma, Vineet Ahuja, and Jaishree Paul. "Real-time analysis of gut flora in *Entamoeba histolytica* infected patients of Northern India." *BMC microbiology* 12, no. 1 (2012): 183.

[74] Kaur, Ravinder, Deepti Rawat, Manish Kakkar, Beena Uppal, and V. K. Sharma. "Intestinal parasites in children with diarrhea in Delhi, India." *Southeast Asian journal of tropical medicine and public health* 33, no. 4 (2002): 725-729.

[75] Manochitra, Kumar, Shashiraja Padukone, Philips A. Selvaratthinam, and Subhash Chandra Parija. "Prevalence of intestinal parasites among patients attending a tertiary care centre in South India." *Int J Curr Microbiol App Sci* 5, no. 9 (2016): 190-7.

[76] Pestehchian, Nader, Mahnaz Nazary, Ali Haghighi, Mansour Salehi, and Hosinali Yosefi. "Frequency of *Entamoeba histolytica* and *Entamoeba dispar* prevalence among patients with gastrointestinal complaints in Chelgerd city, southwest of Iran." *Journal of research in medical sciences: the official journal of Isfahan University of Medical Sciences* 16, no. 11 (2011): 1436.

[77] Bahrami, Fares, Ali Haghighi, Ghasem Zamini, Mohammad Bagher Khadem-Erfan, and Eznolla Azargashb. "Prevalence and associated risk factors of intestinal parasitic infections in Kurdistan province, northwest Iran." *Cogent Medicine* 5, no. 1 (2018): 1503777.

[78] Oyibo, Wellington Aghoghovwia, Anthonia Azuka Ike, O. O. Oladosu, O. T. Ojuromi, A. O. Sanyaolu, and A. O. Olugbenga Olusanya. "Prevalence of *Entamoeba histolytica/Entamoeba dispar* among patients with gastro intestinal complaints in Lagos, Nigeria." *Arch Biomed Sci Health* 1 (2013): 10-9.

[79] Zeb, Aurang, Asma Waheed Qureshi, Luqman Khan, Abu Mansoor, Sami Ullah, and Muhammad Irfan. "Prevalence of *Entamoeba histolytica* in district Buner Khyber Pakhtunkhwa, Pakistan." *J. Entomol. Zool. Stud.* 6 (2018): 185-188.

[80] Ngui, Romano, Nur-Amirah Hassan, Nor Muhammad Soffyan Nordin, Norashikin Mohd-Shaharuddin, Li Yen Chang, Cindy Shuan Ju Teh, Kek Heng Chua, Boon Pin Kee, See Ziau Hoe, and Yvonne Ai Lian Lim. "Copro-molecular study of *Entamoeba* infection among the indigenous community in Malaysia: A first report on the species-specific prevalence of *Entamoeba* in dogs." *Acta tropica* 204 (2020): 105334.

[81] Ngui, Romano, Nur-Amirah Hassan, Nor Muhammad Soffyan Nordin, Norashikin Mohd-Shaharuddin, Li Yen Chang, Cindy Shuan Ju Teh, Kek Heng Chua, Boon Pin Kee, See Ziau Hoe, and Yvonne Ai Lian Lim. "Copro-molecular study of *Entamoeba* infection among the indigenous community in Malaysia: A first report on the species-specific prevalence of *Entamoeba* in dogs." *Acta tropica* 204 (2020): 105334.

[82] Feng, Meng, Kishor Pandey, Tetsuo Yanagi, Ting Wang, Chaturong Putaporntip, Somchai Jongwutiwes, Xunjia Cheng, Jeevan B. Sherchand, Basu Dev Pandey, and Hiroshi Tachibana. "Prevalence and genotypic diversity of *Entamoeba* species in inhabitants in Kathmandu, Nepal." *Parasitology research* 117, no. 8 (2018): 2467-2472.

[83] Abbas, Mushtak Talib, Fahmi Yousef Khan, Saif A. Muhsin, Baidaa Al-Dehwe, Mohamed Abukamar, and Abdel-Naser Elzouki. "Epidemiology, clinical features and outcome of liver abscess: a single reference center experience in Qatar." *Oman medical journal* 29, no. 4 (2014): 260.

[84] Rasti, Sima, Mohsen Arbabi, and Hossein Hooshyar. "High Prevalence of *Entamoeba histolytica/Entamoeba dispar* and *Enterobius vermicularis* Among Elderly and Mentally Retarded Residence in Golabchi Center, Kashan, Iran, 2006-2007." (2012): 585-589.

[85] Samie, Amidou, Leah Mahlaule, Peter Mbati, Tomoyoshi Nozaki, and Ali ElBakri. "Prevalence and distribution of *Entamoeba* species in a rural community in northern South Africa." *Food and Waterborne Parasitology* 18 (2020): e00076.

[86] Othman, Nurulhasanah, Zeehaida Mohamed, Jaco J. Verweij, Lim Boon Huat, Alfonso Olivos-Garcia, Chen Yeng, and Rahmah Noordin. "Application of real-time polymerase chain reaction in detection of *Entamoeba histolytica* in pus aspirates of liver abscess patients." *Foodborne pathogens and disease* 7, no. 6 (2010): 637-641.

[87] Amaechi, A. C., C. C. Ohaeri, O. M. Ukpai, P. C. Nwachukwu, and U. K. Ukoha. "Prevalence of *Entamoeba histolytica* among Primary School children in Ukwa West Local Government Area, Abia State, South East, Nigeria." *The Bioscientist Journal* 2, no. 1 (2014): 1-7.

[88] Siddiqua, Tahmina. *Prevalence of Entamoeba histolytica and Giardia lamblia infection among diabetic and non diabetic patients of Bangladesh.* PhD diss., University of Dhaka, 2016.

[89] Samie, A., L. J. Barrett, P. O. Bessong, J. N. Ramalivhana, L. G. Mavhandu, M. Njayou, and R. L. Guerrant. "Seroprevalence of *Entamoeba histolytica* in the context of HIV and AIDS: the case of Vhembe district, in South Africa's Limpopo province." *Annals of Tropical Medicine & Parasitology* 104, no. 1 (2010): 55-63.

[90] Al-Hindi, Adnan, Amira A. Redwan, Ghada O. El-egla, Razan R. Abu Qassem, and Ayed Alshammari. "Prevalence of intestinal parasitic infections among university female students, Gaza, Palestine." *Avicenna Journal of Medicine* 9, no. 4 (2019): 143.

[91] Aguayo-Patrón, Sandra, Reyna Castillo-Fimbres, Luis Quihui-Cota, and Ana María Calderón de la Barca. "Use of real-time polymerase chain reaction to identify *Entamoeba histolytica* in schoolchildren

from northwest Mexico." *The Journal of Infection in Developing Countries* 11, no. 10 (2017): 800-805.

[92] Guevara, Ángel, Yosselin Vicuña, Denisse Costales, Sandra Vivero, Mariella Anselmi, Zeno Bisoffi, and Fabio Formenti. "Use of real-time polymerase chain reaction to differentiate between pathogenic *Entamoeba histolytica* and the nonpathogenic *Entamoeba dispar* in Ecuador." *The American journal of tropical medicine and hygiene* 100, no. 1 (2019): 81-82.

[93] Yimer, Mulat, Yohannes Zenebe, Wondemagegn Mulu, Bayeh Abera, and José M. Saugar. "Molecular prevalence of *Entamoeba histolytica*/dispar infection among patients attending four health centres in north-west Ethiopia." *Tropical doctor* 47, no. 1 (2017): 11-15.

[94] Akinsanya, B., A. Babatunde, M. Olasanmi, and A. A. Adedotun. "Parasitic infections and risk factors associated with Amoebiasis among pregnant women attending antenatal clinics in primary health care centres in Lagos Mainland, Lagos, Nigeria." *Nigerian Annals of Pure and Applied Sciences* 1 (2018): 52-60.

[95] Deere, Jessica R., Michele B. Parsons, Elizabeth V. Lonsdorf, Iddi Lipende, Shadrack Kamenya, D. Anthony Collins, Dominic A. Travis, and Thomas R. Gillespie. "*Entamoeba histolytica* infection in humans, chimpanzees and baboons in the Greater Gombe Ecosystem, Tanzania." *Parasitology* 146, no. 9 (2019): 1116-1122.

[96] Kyany'a, Cecilia, Fredrick Eyase, Elizabeth Odundo, Erick Kipkirui, Nancy Kipkemoi, Ronald Kirera, Cliff Philip et al. "First report of *Entamoeba moshkovskii* in human stool samples from symptomatic and asymptomatic participants in Kenya." *Tropical Diseases, Travel Medicine and Vaccines* 5, no. 1 (2019): 1-6.

[97] Bahrami, Fares, Ali Haghighi, Ghasem Zamini, and Mohammadbagher Khademerfan. "Differential detection of *Entamoeba histolytica*, *Entamoeba dispar* and *Entamoeba moshkovskii* in faecal samples using nested multiplex PCR in west of Iran." *Epidemiology & Infection* 147 (2019).

[98] Anuar, Tengku Shahrul, Hesham M. Al-Mekhlafi, Mohamed Kamel Abdul Ghani, Siti Nor Azreen, Fatmah Md Salleh, Nuraffini Ghazali, Mekadina Bernadus, and Norhayati Moktar. "First molecular identification of *Entamoeba moshkovskii* in Malaysia." *Parasitology* 139, no. 12 (2012): 1521.

[99] Healy, G. R. "Intestinal and urogenital protozoa." *Manual of clinical microbiology* (1995): 1204-1228.

[100] Garcia, Lynne Shore. *Diagnostic medical parasitology*. Vol. 49. John Wiley & Sons, 2016.

[101] Reed, S. L. Clinical manifestations and diagnosis. In: J. I. Ravdin (ed.), *Amebiasis*. Imperial College Press, London, United Kingdom (2000): 113-126.

[102] Guerrant, Richard L. "The global problem of amebiasis: current status, research needs, and opportunities for progress." *Rev Infect Dis* 8 (1986): 218-227.

[103] Gatti, Simonetta, Giovanni Swierczynski, Francisco Robinson, Mariella Anselmi, Javier Corrales, Juan Moreira, Gregorio Montalvo et al. "Amebic infections due to the *Entamoeba histolytica-Entamoeba dispar* complex: a study of the incidence in a remote rural area of Ecuador." *The American journal of tropical medicine and hygiene* 67, no. 1 (2002): 123-127.

[104] Gilchrist, Carol A., Sarah E. Petri, Brittany N. Schneider, Daniel J. Reichman, Nona Jiang, Sharmin Begum, Koji Watanabe et al. "Role of the gut microbiota of children in diarrhea due to the protozoan parasite *Entamoeba histolytica*." *The Journal of infectious diseases* 213, no. 10 (2016): 1579-1585.

[105] Ghosh, Swagata, Jay Padalia, and Shannon Moonah. "Tissue destruction caused by *Entamoeba histolytica* parasite: cell death, inflammation, invasion, and the gut microbiome." *Current clinical microbiology reports* 6, no. 1 (2019): 51-57.

[106] Gilchrist, Carol A., Brittany N. Schneider, Masud Alam, Jeffrey R. Donowitz, Mamun Kabir, William A. Petri Jr, and Rashidul Haque. "Gut microbiome change prior to the onset of amebiasis." In:

American Journal of Tropical Medicine and Hygiene, vol. 95, no. 5, pp. 172-172.

[107] Stanley Jr, Samuel L. "Amoebiasis." *The lancet* 361, no. 9362 (2003): 1025-1034.

[108] Wuerz, Terry, Jennifer B. Kane, Andrea K. Boggild, Sigmund Krajden, Jay S. Keystone, Milan Fuksa, Kevin C. Kain, Ralph Warren, John Kempston, and Joe Anderson. "A review of amoebic liver abscess for clinicians in a nonendemic setting." *Canadian Journal of Gastroenterology* 26 (2012).

[109] Abd-Alla, Mohamed D., T. G. Jackson, and Jonathan I. Ravdin. "Serum IgM antibody response to the galactose-inhibitable adherence lectin of *Entamoeba histolytica*." *The American journal of tropical medicine and hygiene* 59, no. 3 (1998): 431-434.

[110] Gathiram, T. F. H. G. "A longitudinal study of asymptomatic carriers of pathogenic zymodemes of *Entamoeba histolytica*." *South African Medical Journal* 72, no. 10 (1987): 669-672.

[111] Irusen, E. M., T. F. H. G. Jackson, and A. E. Simjee. "Asymptomatic intestinal colonization by pathogenic *Entamoeba histolytica* in amebic liver abscess: prevalence, response to therapy, and pathogenic potential." *Clinical infectious diseases* 14, no. 4 (1992): 889-893.

[112] Shirley, Debbie-Ann, and Shannon Moonah. "Fulminant amebic colitis after corticosteroid therapy: a systematic review." *PLoS neglected tropical diseases* 10, no. 7 (2016): e0004879.

[113] Tanyuksel, M., & Petri, W. A. (2003). Laboratory diagnosis of amebiasis. *Clinical microbiology reviews*, *16*(4), 713-729.

[114] Aristizábal, Humberto, Jairo Acevedo, and Mario Botero. "Fulminant amebic colitis." *World journal of surgery* 15, no. 2 (1991): 216-221.

[115] Hsu, Y. B., F. M. Chen, P. H. Lee, S. C. Yu, K. M. Chen, Y. T. Yao, and H. C. Hsu. "Fulminant amebiasis: a clinical evaluation." *Hepato-gastroenterology* 42, no. 2 (1995): 109-112.

[116] Rennert, Wolfgang, and Charlotte Ray. "Fulminant amebic colitis in a ten-day-old infant." *The Pediatric infectious disease journal* 19, no. 11 (2000): 1111-1112.

[117] Ortiz-Castillo, Fátima, Luis Enrique Salinas-Aragón, Martín Sánchez-Aguilar, J. Humberto Tapia-Pérez, Martín Sánchez-Reyna, Mauricio Pierdant-Perez, José Juan Sánchez-Rodríguez, and Juan Francisco Hernández-Sierra. "Amoebic toxic colitis: analysis of factors related to mortality." *Pathogens and global health* 106, no. 4 (2012): 245-248.

[118] Takahashi, Takeshi, Armando Gamboa-Dominguez, Tito JM Gomez-Mendez, Jose Maria Remes, Veronica Rembis, Dolores Martinez-Gonzalez, Julieta Gutierrez-Saldivar, Julio Cesar Morales, Jorge Granados, and Juan Sierra-Madero. "Fulminant amebic colitis." *Diseases of the colon & rectum* 40, no. 11 (1997): 1362-1367.

[119] Moorchung, Nikhil, Vikram Singh, Vadalamannati Srinivas, Shyam Sunder Jaiswal, and Gangandeep Singh. "Caecal amebic colitis mimicking obstructing right sided colonic carcinoma with liver metastases: a rare case." *Journal of cancer research and therapeutics* 10, no. 2 (2014): 440.

[120] Sinharay, R., G. K. Atkin, W. Mohamid, and N. Reay-Jones. "Caecal amoebic colitis mimicking a colorectal cancer." *Journal of Surgical Case Reports* 2011, no. 11 (2011): 1-1.

[121] Misra, Sri Prakash, Vatsala Misra, and Manisha Dwivedi. "Ileocecal masses in patients with amebic liver abscess: etiology and management." *World journal of gastroenterology: WJG* 12, no. 12 (2006): 1933.

[122] Magaña-García, Mario, and Antonio Arista-Viveros. "Cutaneous amebiasis in children." *Pediatric dermatology* 10, no. 4 (1993): 352-355.

[123] Mhlanga, Bekithemba R., Leo O. Lanoie, H. Jason Norris, Ernest E. Lack, and Daniel H. Connor. "Amebiasis complicating carcinomas: a diagnostic dilemma." *The American journal of tropical medicine and hygiene* 46, no. 6 (1992): 759-764.

[124] Lysy, Joseph, Joseph Zimmerman, Yoav Sherman, Rivka Feigin, and Moshe Ligumsky. "Crohn's Colitis Complicated by Superimposed Invasive Amebic Colitis." *American Journal of Gastroenterology* 86, no. 8 (1991).

[125] Prakash, V., and S. S. Bhimji. *Amebic Liver*. StatPearls Publishing. (2017).

[126] Singh, Aradhana, Tuhina Banerjee, Raju Kumar, and Sunit Kumar Shukla. "Prevalence of cases of amebic liver abscess in a tertiary care centre in India: A study on risk factors, associated microflora and strain variation of *Entamoeba histolytica*." *PloS one* 14, no. 4 (2019): e0214880.

[127] Haque, Rashidul, Christopher D. Huston, Molly Hughes, Eric Houpt, and William A. Petri Jr. "*Entamoeba histolytica*." *N Engl J Med* 348 (2003): 1565-73.

[128] Lotter, Hannelore, Thomas Jacobs, Iris Gaworski, and Egbert Tannich. "Sexual dimorphism in the control of amebic liver abscess in a mouse model of disease." *Infection and immunity* 74, no. 1 (2006): 118-124.

[129] Maltz, Gary, and C. Michael Knauer. "Amebic liver abscess: a 15-year experience." *American Journal of Gastroenterology* 86, no. 6 (1991).

[130] Thompson Jr, Jesse E., and Alice J. Glasser. "Amebic abscess of the liver: Diagnostic features." *Journal of Clinical Gastroenterology* 8, no. 5 (1986): 550-554.

[131] Lodhi, S., A. R. Sarwari, M. Muzammil, A. Salam, and R. A. Smego. "Features distinguishing amoebic from pyogenic liver abscess: a review of 577 adult cases." *Tropical Medicine & International Health* 9, no. 6 (2004): 718-723.

[132] Pineda, Erika, and Doranda Perdomo. "*Entamoeba histolytica* under oxidative stress: what countermeasure mechanisms are in place?." *Cells* 6, no. 4 (2017): 44.

[133] Adeyemo, Adebayo O., and Adebimpe Aderounmu. "Intrathoracic complications of amoebic liver abscess." *Journal of the Royal Society of Medicine* 77, no. 1 (1984): 17.

[134] Rao, Suchitra, Shahram Solaymani-Mohammadi, William A. Petri Jr, and Sarah K. Parker. "Hepatic amebiasis: a reminder of the complications." *Current opinion in pediatrics* 21, no. 1 (2009): 145.

[135] Nunes, Maria Carmo P., Milton Henriques Guimarães Júnior, Adriana Costa Diamantino, Claudio Leo Gelape, and Teresa Cristina Abreu Ferrari. "Cardiac manifestations of parasitic diseases." *Heart* 103, no. 9 (2017): 651-658.

[136] Kantor, Micaella, Anarella Abrantes, Andrea Estevez, Alan Schiller, Jose Torrent, Jose Gascon, Robert Hernandez, and Christopher Ochner. "*Entamoeba histolytica*: updates in clinical manifestation, pathogenesis, and vaccine development." *Canadian Journal of Gastroenterology and Hepatology* 2018 (2018).

[137] Meyer, Ana-Claire, and Gretchen L. Birbeck. "Parasitic Infections." *Neurobiology of disease* (2007): 453-472.

[138] Hara, Akira, Yoshinobu Hirose, Hideki Mori, Humihiko Iwao, Tatsuo Kato, and Yasuhiro Kusuhara. "Cytopathologic and genetic diagnosis of pulmonary amebiasis: a case report." *Acta cytologica* 48, no. 4 (2004): 547-550.

[139] Zhu, Hailong, Xiangyang Min, Shuai Li, Meng Feng, Guofeng Zhang, and Xianghua Yi. "Amebic lung abscess with coexisting lung adenocarcinoma: an unusual case of amebiasis." *International Journal of Clinical and Experimental Pathology* 7, no. 11 (2014): 8251.

[140] Shenoy, Vishnu Prasad, Shashidhar Vishwanath, Bairy Indira, and G. Rodrigues. "Hepato-pulmonary amebiasis: a case report." *Brazilian Journal of Infectious Diseases* 14, no. 4 (2010): 372-373.

[141] Chalhoub, Sanaa, and Zeina Kanafani. "*Entamoeba histolytica* pleuropulmonary infection: Case report and review of the literature." *Lebanese Medical Journal* 103, no. 365 (2012): 1-3.

[142] Rachid, H., A. Yazidi Alaoui, F. Loudadssi, M. Biaze El, A. Bakhatar, N. Yassine, A. Meziane El, and A. Bahlaoui. "Amoebic infections of the lung and pleura." *Revue des maladies respiratoires* 22, no. 6 Pt 1 (2005): 1035-1037.

[143] Walsh, Julia A. "Problems in recognition and diagnosis of amebiasis: estimation of the global magnitude of morbidity and mortality." *Reviews of infectious diseases* 8, no. 2 (1986): 228-238.

[144] Ohnishi, K., Y. Kato, A. Imamura, M. Fukayama, T. Tsunoda, Y. Sakaue, M. Sakamoto, and H. Sagara. "Present characteristics of symptomatic *Entamoeba histolytica* infection in the big cities of Japan." *Epidemiology & Infection* 132, no. 1 (2004): 57-60.

[145] Huston, Christopher D., Rashidul Haque, and William A. Petri. "Molecular-based diagnosis of *Entamoeba histolytica* infection." *Expert reviews in molecular medicine* 1, no. 9 (1999): 1-11.

[146] Kusuhara, Yasuhiro. "Effectiveness of BD CytoRichTM blue method as a quick Kohn's one-step staining method for *Entamoeba histolytica* trophozoites and Giardia intestinalis cysts." *Int J Anal Bio-Sci Vol* 8, no. 1 (2020).

[147] Proctor, Eileen M. "Laboratory diagnosis of amebiasis." *Clinics in laboratory medicine* 11, no. 4 (1991): 829-859.

[148] Haque, Rashidul, Laurie M. Neville, Pauline Hahn, and W. A. Petri. "Rapid diagnosis of *Entamoeba* infection by using *Entamoeba* and *Entamoeba histolytica* stool antigen detection kits." *Journal of Clinical Microbiology* 33, no. 10 (1995): 2558-2561.

[149] Gonzalez-Ruiz, A., R. Haque, A. Aguirre, G. Castanon, A. Hall, F. Guhl, G. Ruiz-Palacios, M. A. Miles, and D. C. Warhurst. "Value of microscopy in the diagnosis of dysentery associated with invasive *Entamoeba histolytica*." *Journal of clinical pathology* 47, no. 3 (1994): 236-239.

[150] Clark, C. Graham, and Louis S. Diamond. "Methods for cultivation of luminal parasitic protists of clinical importance." *Clinical microbiology reviews* 15, no. 3 (2002): 329-341.

[151] Diamond, Louis S. "Axenic cultivation of *Entamoeba histolytica*." *Science* 134, no. 3475 (1961): 336-337.

[152] Diamond, Louis S., Dan R. Harlow, and Carol C. Cunnick. "A new medium for the axenic cultivation of *Entamoeba histolytica* and other *Entamoeba*." *Transactions of the Royal Society of Tropical Medicine and Hygiene* 72, no. 4 (1978): 431-432.

[153] Diamond, Louis S., C. Graham Clark, and Carol C. Cunnick. "YI-S, a casein-free medium for axenic cultivation of *Entamoeba histolytica*, related *Entamoeba*, Giardia intestinalis and Trichomonas vaginalis." *Journal of Eukaryotic Microbiology* 42, no. 3 (1995): 277-278.

[154] Clark, C. Graham. "Axenic cultivation of *Entamoeba dispar* Brumpt 1925, *Entamoeba insolita* Geiman and Wichterman 1937 and *Entamoeba ranarum* Grassi 1879." *Journal of Eukaryotic Microbiology* 42, no. 5 (1995): 590-593.

[155] Kobayashi, Seiki, Eiko Imai, Hiroshi Tachibana, Tatsushi Fujiwara, and Tsutomu Takeuchi. "*Entamoeba dispar*: Cultivation with Sterilized Crithidia fasciculata 1." *Journal of Eukaryotic Microbiology* 45, no. 2 (1998): 3S-8S.

[156] Diamond, Louis S. "A new liquid medium for xenic cultivation of *Entamoeba histolytica* and other lumen-dwelling protozoa." *Journal of Parasitology* 68, no. 5 (1982): 958-959.

[157] Sargeaunt, P. G. "The reliability of *Entamoeba histolytica* zymodemes in clinical diagnosis." *Parasitology Today* 3, no. 2 (1987): 40-43.

[158] Blanc, D., and P. G. Sargeaunt. "*Entamoeba histolytica* zymodemes: exhibition of γ and δ bands only of glucose phosphate isomerase and phosphoglucomutase may be influenced by starch content in the medium." *Experimental parasitology* 72, no. 1 (1991): 87-90.

[159] Petri Jr, William A., and Upinder Singh. "Diagnosis and management of amebiasis." *Clinical infectious diseases* 29, no. 5 (1999): 1117-1125.

[160] Mirelman, D., Y. Nuchamowitz, and T. Stolarsky. "Comparison of use of enzyme-linked immunosorbent assay-based kits and PCR amplification of rRNA genes for simultaneous detection of *Entamoeba histolytica* and *E. dispar*." *Journal of Clinical Microbiology* 35, no. 9 (1997): 2405-2407.

[161] Wonsit, Ratri, Nitaya Thammapalerd, Savanat Tharavanij, Prayong Radomyos, and Danai Bunnag. "Enzyme-linked immunosorbent assay based on monoclonal and polyclonal antibodies for the

detection of *Entamoeba histolytica* antigens in faecal specimens." *Transactions of the Royal Society of Tropical Medicine and Hygiene* 86, no. 2 (1992): 166-169.

[162] Haque, Rashidul, A. S. G. Faruque, Pauline Hahn, David M. Lyerly, and William A. Petri Jr. "*Entamoeba histolytica* and *Entamoeba dispar* infection in children in Bangladesh." *Journal of Infectious Diseases* 175, no. 3 (1997): 734-736.

[163] Khairnar, Krishna, and Subhash Chandra Parija. "Detection of *Entamoeba histolytica* DNA in the saliva of amoebic liver abscess patients who received prior treatment with metronidazole." *Journal of health, population, and nutrition* 26, no. 4 (2008): 418.

[164] Caballero-Salcedo, Arturo, Monica Viveros-Rogel, Benito Salvatierra, Roberto Tapia-Conyer, Jaime Sepulveda-Amor, Gonzalo Gutierrez, and Librado Ortiz-Ortiz. "Seroepidemiology of amebiasis in Mexico." *The American journal of tropical medicine and hygiene* 50, no. 4 (1994): 412-419.

[165] Calderaro, Adriana, Chiara Gorrini, Simona Bommezzadri, Giovanna Piccolo, Giuseppe Dettori, and Carlo Chezzi. "*Entamoeba histolytica* and *Entamoeba dispar*: comparison of two PCR assays for diagnosis in a non-endemic setting." *Transactions of the Royal Society of Tropical Medicine and Hygiene* 100, no. 5 (2006): 450-457.

[166] Hamzah, Zulhainan, Songsak Petmitr, Mathirut Mungthin, Saovanee Leelayoova, and Porntip Chavalitshewinkoon-Petmitr. "Differential detection of *Entamoeba histolytica*, *Entamoeba dispar*, and *Entamoeba moshkovskii* by a single-round PCR assay." *Journal of Clinical Microbiology* 44, no. 9 (2006): 3196-3200.

[167] ElBakri, Ali, Amidou Samie, Sinda Ezzedine, and Raed Abu Odeh. "Differential detection of *Entamoeba histolytica*, *Entamoeba dispar* and *Entamoeba moshkovskii* in fecal samples by nested PCR in the United Arab Emirates (UAE)." *Acta Parasitologica* 58, no. 2 (2013): 185-190.

[168] Troll, Heike, Hanspeter Marti, and Niklaus Weiss. "Simple differential detection of *Entamoeba histolytica* and *Entamoeba*

dispar in fresh stool specimens by sodium acetate-acetic acid-formalin concentration and PCR." *Journal of Clinical Microbiology* 35, no. 7 (1997): 1701-1705.

[169] Stanley, Samuel L., Annette Becker, Cynthia Kunz-Jenkins, Lynne Foster, and Ellen Li. "Cloning and expression of a membrane antigen of *Entamoeba histolytica* possessing multiple tandem repeats." *Proceedings of the National Academy of Sciences* 87, no. 13 (1990): 4976-4980.

[170] de la Vega, Humberto, Charles A. Specht, Carlos E. Semino, Phillips W. Robbins, Daniel Eichinger, Daniel Caplivski, Sudip Ghosh, and John Samuelson. "Cloning and expression of chitinases of *Entamoebae*." *Molecular and biochemical parasitology* 85, no. 2 (1997): 139-147.

[171] Zindrou, Sherwan, Esther Orozco, Ewert Linder, Aleyda Téllez, and Anders Björkman. "Specific detection of *Entamoeba histolytica* DNA by hemolysin gene targeted PCR." *Acta tropica* 78, no. 2 (2001): 117-125.

[172] Tannich, E., and G. D. Burchard. "Differentiation of pathogenic from nonpathogenic *Entamoeba histolytica* by restriction fragment analysis of a single gene amplified in vitro." *Journal of Clinical Microbiology* 29, no. 2 (1991): 250-255.

[173] Tachibana, Hiroshi, Seiki Kobayashi, Masataka Takekoshi, and Seiji Ihara. "Distinguishing pathogenic isolates of *Entamoeba histolytica* by polymerase chain reaction." *Journal of Infectious Diseases* 164, no. 4 (1991): 825-826.

[174] Khairnar, Krishna, and Subhash C. Parija. "A novel nested multiplex polymerase chain reaction (PCR) assay for differential detection of *Entamoeba histolytica*, *E. moshkovskii* and *E. dispar* DNA in stool samples." *BMC microbiology* 7, no. 1 (2007): 47.

[175] Blessmann, Joerg, Heidrun Buss, Phuong A. Ton Nu, Binh T. Dinh, Quynh T. Viet Ngo, An Le Van, Mohamed D. Abd Alla, Terry FHG Jackson, Jonathan I. Ravdin, and Egbert Tannich. "Real-time PCR for detection and differentiation of *Entamoeba histolytica* and

Entamoeba dispar in fecal samples." *Journal of Clinical Microbiology* 40, no. 12 (2002): 4413-4417.

[176] Klein, Dieter. "Quantification using real-time PCR technology: applications and limitations." *Trends in molecular medicine* 8, no. 6 (2002): 257-260.

[177] Kebede, A., J. J. Verweij, T. Endeshaw, T. Messele, G. Tasew, B. Petros, and A. M. Polderman. "The use of real-time PCR to identify *Entamoeba histolytica* and *E. dispar* infections in prisoners and primary-school children in Ethiopia." *Annals of Tropical Medicine & Parasitology* 98, no. 1 (2004): 43-48.

[178] Verweij, Jaco J., Roy A. Blangé, Kate Templeton, Janke Schinkel, Eric AT Brienen, Marianne AA van Rooyen, Lisette van Lieshout, and Anton M. Polderman. "Simultaneous detection of *Entamoeba histolytica*, Giardia lamblia, and Cryptosporidium parvum in fecal samples by using multiplex real-time PCR." *Journal of clinical microbiology* 42, no. 3 (2004): 1220-1223.

[179] Verweij, Jaco J., Fieke Oostvogel, Eric AT Brienen, Alexis Nang-Beifubah, Juventus Ziem, and Anton M. Polderman. "Prevalence of *Entamoeba histolytica* and *Entamoeba dispar* in northern Ghana." *Tropical Medicine & International Health* 8, no. 12 (2003): 1153-1156.

[180] Qvarnstrom, Yvonne, Cleve James, Maniphet Xayavong, Brian P. Holloway, Govinda S. Visvesvara, Rama Sriram, and Alexandre J. da Silva. "Comparison of real-time PCR protocols for differential laboratory diagnosis of amebiasis." *Journal of clinical microbiology* 43, no. 11 (2005): 5491-5497.

[181] Yeh, Shiou-Hwei, Ching-Yi Tsai, Jia-Horng Kao, Chun-Jen Liu, Ti-Jung Kuo, Ming-Wei Lin, Wen-Ling Huang et al. "Quantification and genotyping of hepatitis B virus in a single reaction by real-time PCR and melting curve analysis." *Journal of hepatology* 41, no. 4 (2004): 659-666.

[182] Njiru, Zablon Kithinji. "Loop-mediated isothermal amplification technology: towards point of care diagnostics." *PLoS Negl Trop Dis* 6, no. 6 (2012): e1572.

[183] Rivera, Windell L., and Vanissa A. Ong. "Development of loop-mediated isothermal amplification for rapid detection of *Entamoeba histolytica*." *Asian Pac J Trop Med* 6, no. 6 (2013): 457-461.

[184] Liang, Shih-Yu, Yun-Hsien Chan, Kan-Tai Hsia, Jing-Lun Lee, Ming-Chu Kuo, Kuo-Yuan Hwa, Chi-Wen Chan et al. "Development of loop-mediated isothermal amplification assay for detection of *Entamoeba histolytica*." *Journal of Clinical Microbiology* 47, no. 6 (2009): 1892-1895.

[185] Mwendwa, Fridah, Cecilia K. Mbae, Johnson Kinyua, Erastus Mulinge, Gitonga Nkanata Mburugu, and Zablon K. Njiru. "Stem loop-mediated isothermal amplification test: comparative analysis with classical LAMP and PCR in detection of *Entamoeba histolytica* in Kenya." *BMC Research Notes* 10, no. 1 (2017): 1-9.

[186] Wang, Zheng, Gary J. Vora, and David A. Stenger. "Detection and genotyping of *Entamoeba histolytica*, *Entamoeba dispar*, Giardia lamblia, and Cryptosporidium parvum by oligonucleotide microarray." *Journal of Clinical Microbiology* 42, no. 7 (2004): 3262-3271.

[187] Shah, Preetam H., Ryan C. MacFarlane, Dhruva Bhattacharya, John C. Matese, Janos Demeter, Suzanne E. Stroup, and Upinder Singh. "Comparative genomic hybridizations of *Entamoeba* strains reveal unique genetic fingerprints that correlate with virulence." *Eukaryotic Cell* 4, no. 3 (2005): 504-515.

[188] Haque, R., Mondal, D., Duggal, P., Kabir, M., Roy, S., Farr, B. M. et al. (2006). *Entamoeba histolytica* infection in children and protection from subsequent amebiasis. Infection and Immunity. 74, pp. 904–909.

[189] Gonzales, Maria Liza M., Leonila F. Dans, and Juliet Sio-Aguilar. "Antiamoebic drugs for treating amoebic colitis." *Cochrane Database of Systematic Reviews* 1 (2019).

[190] Connor, Thomas H., Marie Stoeckel, John Evrard, and Marvin S. Legator. "The contribution of metronidazole and two metabolites to the mutagenic activity detected in urine of treated humans and mice." *Cancer research* 37, no. 2 (1977): 629-633.

[191] Lindmark, Donald G., and Miklós Müller. "Antitrichomonad action, mutagenicity, and reduction of metronidazole and other nitroimidazoles." *Antimicrobial agents and chemotherapy* 10, no. 3 (1976): 476-482.

[192] Rossignol, Jean-François, Samir M. Kabil, Yehia El-Gohary, and Azza M. Younis. "Nitazoxanide in the treatment of amoebiasis." *Transactions of the Royal Society of Tropical Medicine and Hygiene* 101, no. 10 (2007): 1025-1031.

[193] Patel, J. C. "Chloroquine in the treatment of amoebic liver abscess." *British medical journal* 1, no. 4814 (1953): 811.

[194] *Drugs for parasitic infections.* 2nd ed. Abramowicz J, editor. New Rochelle, New York: 2010.

[195] Capparelli, Edmund V., Robin Bricker-Ford, M. John Rogers, James H. McKerrow, and Sharon L. Reed. "Phase I clinical trial results of auranofin, a novel antiparasitic agent." *Antimicrobial agents and chemotherapy* 61, no. 1 (2017).

[196] Mori, Mihoko, Satoshi Tsuge, Wataru Fukasawa, Ghulam Jeelani, Kumiko Nakada-Tsukui, Kenichi Nonaka, Atsuko Matsumoto, Satoshi Ōmura, Tomoyoshi Nozaki, and Kazuro Shiomi. "Discovery of antiamebic compounds that inhibit cysteine synthase from the enteric parasitic protist *Entamoeba histolytica* by screening of microbial secondary metabolites." *Frontiers in cellular and infection microbiology* 8 (2018): 409.

[197] Shahinas, Dea, Anjan Debnath, Christan Benedict, James H. McKerrow, and Dylan R. Pillai. "Heat shock protein 90 inhibitors repurposed against *Entamoeba histolytica*." *Frontiers in microbiology* 6 (2015): 368.

[198] S. M. Shamsuzzaman and Y. Hashiguchi, "Thoracicamebiasis," *Clinics in Chest Medicine*, vol. 23, no. 2, (2002): 479–492.

[199] Ghosh, Jayant Kumar, Sundeep Kumar Goyal, Manas Kumar Behera, Manish Kumar Tripathi, Vinod Kumar Dixit, Ashok Kumar Jain, and Ramchandra Shukla. "Efficacy of aspiration in amebic liver abscess." *Tropical Gastroenterology* 36, no. 4 (2016): 251-255.

[200] Quach, Jeanie, Joëlle St-Pierre, and Kris Chadee. "The future for vaccine development against *Entamoeba histolytica*." *Human vaccines & immunotherapeutics* 10, no. 6 (2014): 1514-1521.

[201] Zhang, Tonghai, Paul R. Cieslak, and SLj Stanley. "Protection of gerbils from amebic liver abscess by immunization with a recombinant *Entamoeba histolytica* antigen." *Infection and immunity* 62, no. 4 (1994): 1166-1170.

In: *Entamoeba*
Editor: Thomas L. Johnson
ISBN: 978-1-53618-506-5

Chapter 2

PATHOGENIC AND NON-PATHOGENIC SPECIES OF *ENTAMOEBA*

***Shadab Miyan Siddiqui*[1], *Abdul Roouf Bhat*[2] and *Fareeda Athar*[3]**
[1]Department of Chemistry, Jamia Millia Islamia, New Delhi, India
[2]Department of Chemistry, Sripratap College, Cluster University, Srinagar, India
[3]Center for Interdisciplinary Research in Basic Science, Jamia Millia Islamia, Jamia Nagar, New Delhi, India

ABSTRACT

The genus *Entamoeba* comprises engrossing protists, which are found in humans, nonhuman primates, other vertebrates and invertebrates. *Entamoeba histolytica*, *Entamoeba dispar*, *Entamoeba gingivalis*, *Entamoeba coli*, *Entamoeba moshkovskii*, *Entamoeba invadens*, *Entamoeba polecki*, *Entamoeba hartmanni*, *Entamoeba suis*, *Entamoeba nuttalli*, *Entamoeba bangladeshi*, *Entamoeba struthionis* and *Entamoeba muris* are important organisms of the genus. Of these, a few species are harmless, while a few are pathogenic. *Entamoeba gingavalis* was the first parasitic amoeba reported in humans, later on some other species have been found in human, of these, *Entamoeba histolytica* is most fatal species that affects more than 10% of the world's population and causes amoebiasis, which is an emerging parasitic complication in the human immunodeficiency virus (HIV)-infected patients and is responsible for approximately 100,000 fatalities annually, making it the second leading cause of death due to parasitic disease. Moreover, the species which are morphologically indistinguishable from *E. histolytica* are of immense importance because these may be confused with *E. histolytica* in diagnostic investigations. In order to control infectious agent, sound knowledge of taxonomy, prevalence, morphology and mode of transmission of the infectious agent is extremely essential, therefore, in this context, the present chapter provides the details of various species belonging to the genus *Entamoeba*.

Keywords: Protozoa, *Entamoeba*, Amoebiasis

1. Introduction and Classification

The genus *Entamoeba* includes several species of unicellular organisms residing in humans, nonhuman primates, other vertebrates, invertebrates and a few of them are free livings species. *Entamoeba* spp. are taxonomically within the kingdom Protista, subkingdom Protozoa, phylum Sarcomastigophora, subphylum Sarcodina, class Lobosea, order Amoebida and family Entamoebidae (Silberman et al., 1999). *Entamoebae* were previously considered to be the most primitive extant eukaryotic group-Archezoa (Baldauf et al. 2008). Due to the lack of the typical eukaryotic "mitochondria" organelles, *Entamoeba* was deemed a living relic, a paradigm for the earliest eukaryotic cell (Clark et al., 2000). The genus *Entamoeba* was named in 1875 by Casagrandi and Barbagallo and *Entamoeba gingavalis* was the first amoeba which was found in humans (Al-Saeed et al., 2003), however, later on some other species have been found in human. *Entamoeba histolytica, Entamoeba dispar, Entamoeba gingivalis, Entamoeba coli, Entamoeba moshkovskii, Entamoeba invadens, Entamoeba polecki, Entamoeba hartmanni, Entamoeba suis, Entamoeba nuttalli, Entamoeba bangladeshi, Entamoeba struthionis* and *Entamoeba muris* are important organisms of the genus *Entamoeba.*

In the first quarter of the 20th century, dozens of species names were assigned to amoebae of the genus *Entamoeba* that were often based largely on the parasite being found in a new host (Dobell, 1919). The simplest morphological feature to use in *Entamoeba* species identification is the number of nuclei per cyst. A previous phylogenetic study of *Entamoeba* species suggested that this morphological feature reflected the phylogenetic relationships among organisms, with the species producing cysts with different numbers of nuclei forming distinct clades (Silberman et al., 1999).

The organisms of the genus *Entamoeba* have a direct life cycle, usually the cyst serving as infective stage from one host to another; however the species like *E. gingivalis* do not form cysts. The presence/absence of mature cyst and their morphology have been used to classify the species of the genus into groups. Species of the genus *Entamoeba* have been divided to five groups based on the number of nuclei willing in mature cyst by clades (Levin et al. 1985). These groups are as follows:

A. Species without cyst or *E. gingivalis* –like group;
B. Species with one nucleate mature cyst or *E. bovis*-like group;
C. Species with four nucleate mature cyst or *E. histolytica*-like group;
D. Species with eight nucleate mature cyst or *E. coli* -like group;
E. inadequately known species.

The validity of this category was confirmed by using riboprinting method by Clarck and Diamond (Clarck et al. 1997). *Entamoeba histolytica* is the most important species of the genus *Entamoeba*. The species of the genus are not only limited to human but these have broad range of hosts. Apart from the well known human infecting species, several other species that are not found in humans have been identified in non human primates. *Entamoeba hartmanni* is commonly found in non human primates. *Entamoeba dispar* is also widespread in non human primates. *Entamoeba chattoni* which was considered as a subtype of *Entamoeba Polecki*, now has been accepted as a non human primate species. Moreover, *E. histolytica* of human origin has been shown experimentally to be capable of infecting non human primates where it can cause indistinguishable pathology (Haq et al. 1985). Although, all the species are not pathogenic, a few species are harmless yet they are as important as pathogenic species because, some of the non pathogenic species are morphologically identical to pathogenic species of *Entamoeba*. The various species of the genus *Entamoeba* are of immense importance to human as well as the domestic animal health.

2. Various Species of Genus *Entamoeba*

The genus *Entamoeba* comprises several unicellular organism. The cells of *Entamoeba* species contains a nucleus with a small nucleolus (karyosome) and peripheral chromatin. The nucleus has a "bull's eye" pattern due to peripheral chromatin and a small, dense karyosome in the center. The genus *Entamoeba* is well known for *E. histolytica*, which is the pathogenic species and classified as a category priority biodefence pathogen by the National institute of Allergy and infectious Disease (Debnath et al. 2012). It is intriguing to speculate that other *Entamoeba* species which do not apparently cause clinical diseases may still have a large impact on host microflora composition and indirectly affect health and wellness (Everard et al. 2014). Some species of the genus are well known and some of them are not well studied yet. Since the differentiation among the species is extremely important for the diagnosis of individuals at risk and to prevent unnecessary chemotherapy, the knowledge about the various species of the genus, is of immense importance. The various species belonging to the genus *Entamoeba* have been described herein.

2.1. *Entamoeba histolytica* (*E. histolytica*)

Entamoeba histolytica was first described by Fedor Losch in 1875 and the species name *E. histolytica* was first coined by Fritz Schaudinn in 1903. *E. histolytica* has worldwide distribution, with a high incidence in subtropical and tropical regions of developing countries and high prevalence in countries with poor sanitary and socioeconomic conditions. Infection of *E. histolytica* affects more than 10% of the world's population (Marion et al. 2006). *E. histolytica* is well known for its high potential for invading and destroying human tissue, leading to amebic dysentery and a wide range of other invasive diseases, including amebic liver abscess, respiratory tract infections, and cerebral and genitourinary amebiasis (Stanley et al. 2003).

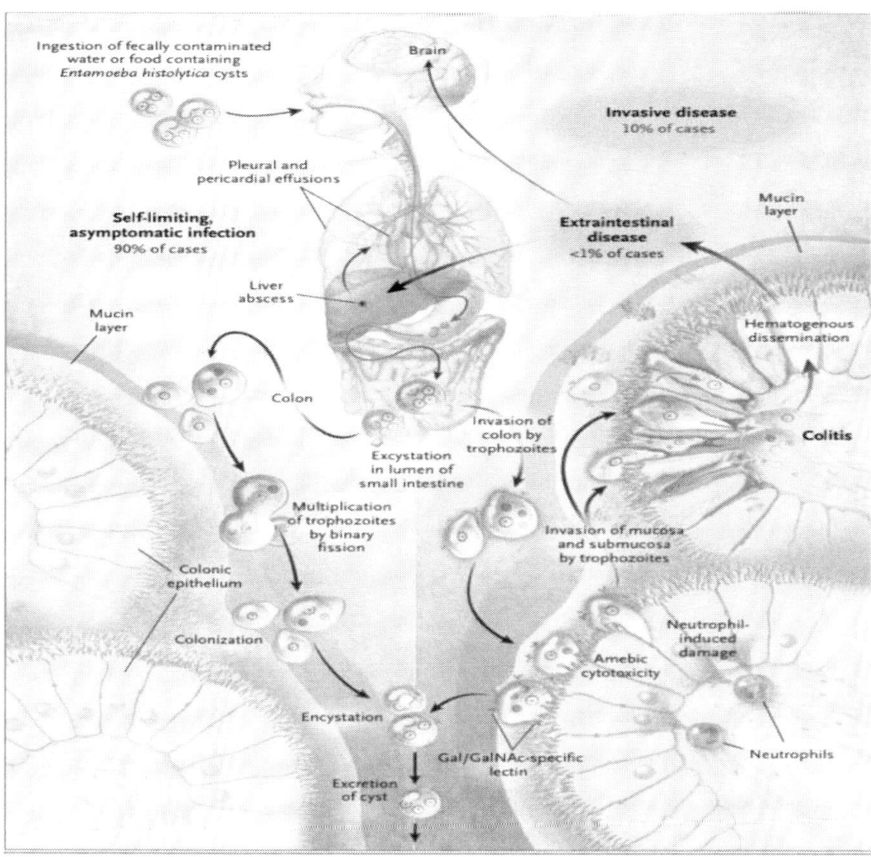

Figure 1. The Life Cycle of *Entamoeba histolytica*.

Entamoeba histolytica is a monogenetic parasite and it exists in two forms: trophozoite and cyst. The trophozoite is the invasive form which is also called as amoeba and cyst is the infective form of the parasite. The life cycle of *E. histolytica* in man begins with ingestion of quardinucleated cysts (Figure 1, Haque et al. 2003).

Cysts are ingested from fecally contaminated food or water or through oral-anal sexual practice. Cyst can tolerate the stomach acidic pH and in the intestine, excysts to trophozoites (Figure 2, Faust et al. 2011). In response to unknown stimuli, amoebae undergo morphological and biochemical changes that leads to the formation of new cysts which are eliminated in the feces, completing the cycle (Martinez-Palomo et al. 1987, Espinosa-Cantellano et al. 2000, Stanley et al. 2003).

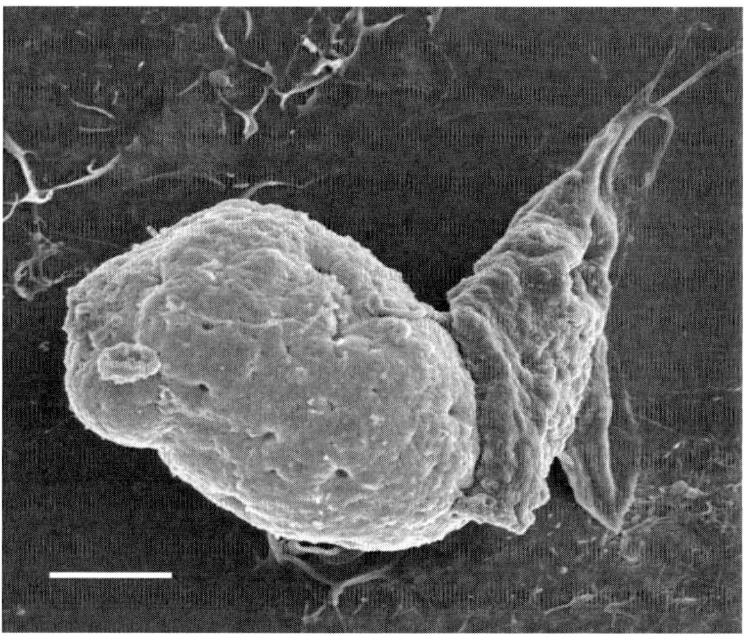

Figure 2. Interaction of virulent *E. histolytica* with human Liver sinusoidal endothelial cells. Scanning Electron Microscopy micrograph of a trophozoite in contact with human liver sinusoida endothelial cells.

E. histolytica thioredoxin reductase and *E. histolytica O*-acetyl serine sulfhydrylase and are two important enzymes which play crucial role in the life cycle of *E. histolytica*. *E. histolytica* thioredoxin reductase maintains intracellular redox balance (Holmgren et al. 2005). The growth and survival of the parasite depend upon the cysteine biosynthetic pathway (Fahey et al. 1984). *O*-acetyl serine sulfhydrylase catalyzes the last step of the cysteine biosynthetic pathway. Cysteine, which is the product of this pathway, is the only antioxidative thiol in *E. histolytica* and plays significant role in maintaining the redox balance in the organism (Arias et al. 2012). *E. histolytica* produces considerable amounts of cysteine proteinases and this class of enzymes is instrumental for *E. histolytica*-induced pathology (Moncada et al. 2003).

2.2. *Entamoeba gingivalis* (*E. gingivalis*)

E. gingivalis is a cosmopolitan species, which was discovered by G. Gros in 1949. *Entamoeba gingivalis* is the first parasitic amoeba reported in humans. It is the only *Entamoeba* species which colonizes the human oral cavity most often found in gingival tissues around the teeth, gums and sometimes tonsils. Feed on epithelial cells of the mouth, bacteria, food debris and other cells available to them (Rousset et al. 1971). It is the only species of *Entamoeba* which can phagocytose nuclear fragments of ingested bacteria, cellular debris, leucocytes and epithelial cells and occasional red blood cells (Ashraf et al. 2012). The parasite is highly enriched in people with periodontitis, a disease leading to alveolar bone destruction and eventually tooth loss. (Kofoid et al. 1929).

In general, most of the species of *Entamoeba*, exist in two forms: cyst and trophozoites. The survival of these *Entamoeba* species is ensured by their encystment in response to environmental changes, permitting the survival in environments exposed to oxygen, like human stools. Encystment does not occur in *E. gingivalis* nevertheless, this parasitic protozoan is found in periodontal pockets, suggesting that low oxygen levels are significant for the survival of the trophozoites.

Entamoeba gingivalis has a large number of pseudopodia that allow it to move quickly. Moreover, it does not have resistant form (cyst), but only trophozoites which are morphologically similar to that of *E. histolytica*. The parasite does not produce cysts, therefore, the transmission is direct from one person to another by kissing or by sharing eating utensils (Figure 3).

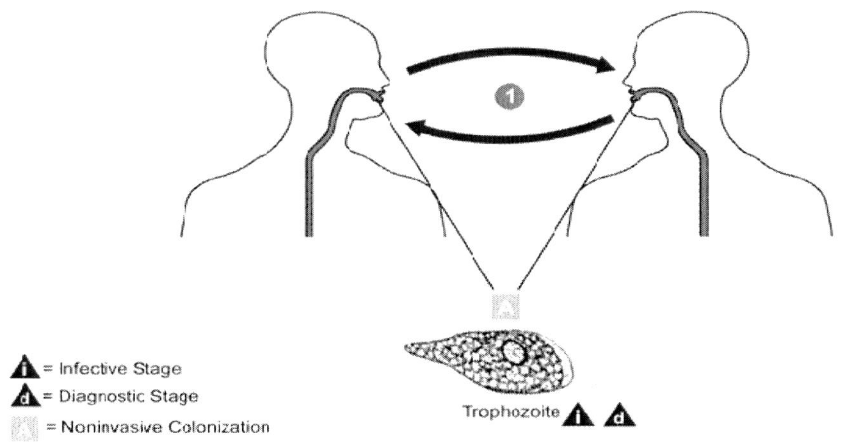

Figure 3. The Life Cycle of *Entamoeba gingivalis*.

2.3. *Entamoeba invadens* (*E. invadens*)

Entamoeba invadens is the most widespread *Entamoeba* that infects reptiles (MacNeill et al., 2002; Kojimoto et al., 2001); however, other *Entamoeba* species, such as *E. terrapinae*, *E. insolita, E. barreti, E. testudines, and E. ranarum*, can also infect reptiles (Garcia et al., 2014). *Entamoeba invadens* causes amoebiasis in reptiles (German et al. 1936). Amoebiasis associated with ulcerative enteritis and hepatitis caused by *Entamoeba invadens* is one of the most common and serious protozoal diseases in lizards, snakes, and some other reptiles (Donaldson et al. 1975).

E. invadens has a direct life cycle with no intermediate host. (German et al. 1936). *The life cycle of E. invadens* exists in two forms: the motile trophozoite that inhabits the host and the cyst which is the infectious form of the parasite. The cyst wall contains chitin, chitosan fibrils and chitin binding proteins. These elements confer resistance to extreme environments before it is ingested by the host (Welter et al. 2017). The cyst releases the trophozoites after oral ingestion by the host (Kojimoto et al., 2001). The trophozoite first adheres to the intestinal mucus and epithelial cells by a Gal/GalNAc-specific lectin (McCoy et al. 1994). Following this intimate contact, pore forming polypeptides called amoebapores are released by the trophozoite. (Espinosa-Cantellano et al. 2000). Trophozoites secondarily enter the mesenteric circulation and invade the liver or other organs, causing hepatic abscesses or other extra-intestinal lesions (Chia et al. 2009).

E. invadens is an important model for encystation process. In most laboratory-based studies of the cyst, this reptilian parasite has been used because *E. histolytica* does not encyst in culture. *E. invadens* can be induced to encyst in axenic laboratory culture, while encystation has not yet been achieved in axenically grown *E. histolytica* trophozoites (Eichinger 1997).

2.4. *Entamoeba dispar* (*E. dispar*)

In 1919, the English parasitologist Clifford Dobell described one species of *Entamoeba* that produced cysts having four nuclei and for this organism he retained name *Entamoeba histolytica*. Later on, in 1993, *E. histolytica* and *E. dispar* were formally accepted as different yet closely related species on the basis of extensive genetic, immunological, and biochemical analyses (Diamond et al. 1993). *E. histolytica* and *E. dispar* are genetically distinct but morphologically identical species. *E. dispar* is ten times more common than *E. histolytica*, however, in developing countries both of the species can be equally prevalent (Haque et al. 1995). However, *E. dispar* resides in the human intestinal lumen, besides it has been found in faces of non-human primates, such as macaque monkeys, organutans and baboons (Feng et al. 2013). There are a few reports that suggest that *E. dispar* may be capable of causing intestinal and extraintestinal disease in humans and experimental animals (Dolabella et al. 2012).

2.5. *Entamoeba moshkovskii* (*E. moshkovskii*)

E. moshkovskii was first identified in sewage in Moscow in 1941 (Shimokawa et al. 2012). Initially it was considered as free living non-pathogenic protozoan species but later on, it has been found that *E. moshkovskii* has the ability to infect human. The first report of human infection with moshkovskii was proposed by Dreyer in 1961. It has been diagnosed in fecal samples of patients with gastrointestinal symptoms in Australia, Bangladesh, India, Iran, Tanzania, and Turkey, suggesting that *E. moshkovskii* act as a causative agent of amoebiasis (Shimokawa et al. 2012). *E. moshkovskii* do not have adhesin but are morphologically similar to *E. histolytica* (McHardy et al. 2014). *E. moshkovskii* has also been detected in cattles, elephants, reptiles and insects (Elsheikha et al. 2018).

2.6. *Entamoeba polecki* (*E. polecki*)

Entamoeba polecki is a parasite of human and nonhuman primates, other mammals, and birds. It was identified in pigs and human for the first time by Prowazek in 1912 (Stensvold et al. 2018). The parasite resides mainly in the colon of the pigs and monkeys (Salaki et al. 1979). However, it occurs worldwide, but symptomatic infection in humans is rare (Desowitz et al. 1986).

E. polecki produces cysts with one nucleus. The cyst measuring 5–18 μm may contain inclusion bodies and abundant fragmented chromatoidal bars. The trophozoite closely resembles those of *E. coli* and *E. histolytica*. Therefore, it is very difficult to discriminate its trophozoite from the trophozoites of *E. coli* and *E. histolytica*: the presence of mature cysts is required for laboratory diagnosis of *E. polecki* (Wiwanitkit et al. 2004). Initially, it was considered as non-pathogenic to human, but later on it was found to be as a possible pathogen, causing diarrhea (Stensvold et al. 2018). It has been implicated in cases of diarrhea and abdominal pain and transmission probably results from transfer of cysts excreted by pigs or monkeys (Desowitz et al. 1986).

2.7. *Entamoeba hartmani* (*E. hartmani*)

Entamoeba hartmani is generally highly endemic in tropical and sub-tropical rural areas (Matsumura et al. 2019). It is a non-pathogenic species which resides in the colon, caecum and large intestine of the human. It is quardi- nucleated cyst clade but it is smaller than *Entamoeba histolytica*, having trophozoites of 3–12 μM m in diameter and the cyst of 10 μM in diameter. *Entamoeba hartmani* is considered non-pathogenic species.

2.8. *Entamoeba coli* (*E. coli*)

Entamoeba coli is a neglected intestinal amoeba, mostly occur in the tropical African countries. It was discovered in India by Lewis in 1870 however its detail description was given by Grassi in 1879. *E. coli* is a monogenetic parasite, residing the lumen of the large intestine of man. Three distinct morphological forms exist in the life cycle-Trophozoite, Pre-cystic stage and Cystic stage. The trophozoites are similar to those of *E. histolytica*, but can be distinguished by their wide and tapered pseudopodia and the cyst are distinguished by the presence of eight nuclei in the mature form. *E. coli* is considered harmless, however, when an individual is infected with *E. coli*, other pathogenic organisms may have been introduced and might cause infection and ailment in that individual. *E. coli* includes two clades which have been termed as ST1 and ST2. Based on sequence divergence, it would be reasonable to consider *E. coli* ST1 and ST2 to be distinct species, however, other than this sequence divergence, there are no other known differences between the two subtypes to date (Elsheikha et al. 2018).

2.9. *Entamoeba bangladeshi* (*E. bangladeshi*)

In 2012, Royer identified a new species of *Entamoeba* from Bangladesh which is known as *Entamoeba bangladeshi*. *Entamoeba bangladeshi* is found in human intestinal lumen and it is very similar to *E. moshkovaskii* and *E. histolytica*. *E. bangladeshi* is morphologically indistinguishable from *E. histolytica* and *E. moshkovskii,* but it has been considered as non- pathogenic species.

2.10. *Entamoeba nuttali* (*E. nuttali*)

Entamoeba nuttali is a potentially pathogenic species of the genus *Entamoeba*. This gastrointestinal pathogen is phylogenetically closest species to *E. histolytica* (Tachibana et al. 2007) and causes invasive amoebiasis in non human primates, that may result in hemorrhagic dysentery, liver abscess or other extraintestinal pathologies, and even death (Loomis et al. 1983).

2.11. *Entamoeba antilocapra* (*E. antilocapra*)

In 1953, Noble reported *E.* anantilocapra in captive pronghorns (*Antilocapra americana*) and named it *Entamoeba antilocaprae*. It resides in the large intestine of Antelope and it has been found to cause Intestinal lesion and bowel inflammation.

2.12. *Entamoeba bovis* (*E. bovis*)

It was first described by Liebetanz in 1905. It resides in the large intestine of cattle and buffalo. There is no evidence of pathogenicity of this organism.

2.13. *Entamoeba apis* (*E. apis*)

Entamoeba apis was reported by Fantham and Porter in 1911. It is a non pathogenic species which resides in the intestine of bee *(Apis mellifica).*

2.14. *Entamoeba struthionis* (*E. struthionis*)

Entamoeba struthionis is the most common protozoa found in ostriches (*Struthio camelus*), which forms mature uninucleated cysts (Ponce-Gordo et al., 2004).

2.15. *Entamoeba gallinarum* (*E. gallinarum*)

Entamoeba gallinarum have been reported to infect birds. It has been found in chicken, turkey, guinea, fowl, duck and goose (Ponce-Gordo et al., 2004).

2.16. *Entamoeba anatis* (*E. anatis*)

Entamoeba anatis is a pathogenic species which has been reported in duck (Ponce-Gordo et al., 2004).

2.17. *Entamoeba muris* (*E. muris*)

Entamoeba muris is a highly contagious enteric protozoan parasite of rats, mice and other redents. *E. muris* is morphologically similar to *Entamoeba coli* (Neal et al., 1950). The mature cyst of both contain eight nuclei and can be distinguished when stained with iodine.

2.18. *Entamoeba suis* (*E. suis*)

Entamoeba suis is a protozoan parasite of pigs. It resides in colon, caecum and large intestine of pig. *E. suis* has been identified as the causative agent of diarrhea in pigs (Komatsu et al. 2019).

2.19. *Entamoeba cobaye (E. cobaye)*

Entamoeba cobaye is a non-pathogenic species, which resides in the large intestine of guinea pig.

2.20. *Entamoeba chattoni* (*E. chattoni*)

Entamoeba chattoni is found in human and monkey. It inhabits in colon, caecum and large intestine. *E. chattoni* is considered non-pathogenic species.

2.21. *Entamoeba citelli (E. citelli)*

Entamoeba citelli inhabits in colon, caecum and large intestine of ground squirrel and it is a non- pathogenic species.

2.22. *Entamoeba criceti* (*E. criceti*)

Entamoeba criceti is a non pathogenic species living in the large intestine of hamster.

2.23. *Entamoeba cuniculi* (*E. cuniculi*)

This non pathogenic species of *Entamoeba* resides in the large bowel of rabbits.

2.24. *Entamoeba chiropteris* (*E. chiropteris*)

Entamoeba chiropteris is a non-pathogenic species living in the large bowel of bats.

2.25. *Entamoeba dipodomysi* (*E. dipodomysi*)

Entamoeba dipodomysi is a non-pathogenic species which is found in the large bowel of kangaroo rats.

2.26. *Entamoeba funambulae* (*E. funambulae*)

It is a non pathogenic organism of the genus *Entamoeba*, which have inhabits in the large intestine of Indian palm squirrel.

2.27. *Entamoeba flaviviridis* (*E. flaviviridis*)

Entamoeba flaviviridis resides in Intestine of lizard and it is a non-pathogenic species.

2.28. *Entamoeba polypodia* (*E. polypodia*)

Entamoeba polypodia is non-pathogenic species and it inhabits in ventricle, intestine and rectum of bug (*Leptocoris trivitlatus*).

2.29. *Entamoeba marmotae* (*E. marmotae*)

Entamoeba marmotae is a non-pathogenic unicellular organism that resides in the large intestine of marmot.

2.30. *Entamoeba wenyoni* (*E. wenyoni*)

E. wenyoni is a non-pathogenic species, which inhabits in the large intestine of goat and camel.

2.31. *Entamoeba ranarum* (*E. ranarum*)

Entamoeba ranarum resides in the large intestine of various species of frogs. Moreover, it can infect reptile and amoebic liver abscess have been reported in reptiles. (Garcia et al., 2014).

2.32. *Entamoeba aulastomi* (*E. aulastomi*)

In, 1912, Noller found *Entamoeba aulastomi* in the hind gut of well nourished horse-leeches. It produces tetra-nucleate cyst which measures 7-11 μm in diameter.

2.33. *Entamoeba lagopodis* (*E. lagopodis*)

Entamoeba lagopodis have been found in the caecum of duck and lagopus. Pathogenicity of this species is unknown.

2.34. *Entamoeba ecuadoriensis* (*E. ecuadoriensis*)

E. ecuadoriensis is a free living species, which have a close resemblance with *E. moshkovskii* and *E. marina*.

2.35. *Entamoeba marina* (*E. marina*)

E. marina was isolated from a sample of tidal flat sediments collected at Iriomote Island, Okinawa, Japan. *E. marina* distantly positioned from two other free living species: *E. moshkovskii* and *E. ecuadoriensis*. However, it is uncertain whether *E. marina* is a free-living species or not because *E. marina* in the sample of tidal flat sediments was derived from small animals (e.g., sand worms, crabs, and shellfish) that living in the sediments or feces of animals visited the tidal flats (Shiratori et al., 2014).

CONCLUSION

The genus *Entamoeba* is diverse, contains several species and wide range of hosts. All the species are not pathogenic but they may be confused with pathogenic species, therefore, in order to avoid unnecessary treatment of the people infected with non-pathogenic *Entamoeba* species, it is essential to discriminate these species from the pathogenic species. In the recent years, the molecular methods have detected considerable diversity within the genus and enable the detection and distinction of the species. However, some organisms have not been accepted as distinct species and may be atypical form or synonym of the known species. A few organisms of the other genus have been misdiagnosed as *Entamoeba* species and many species of *Entamoeba* have not been studied in detail yet, the available informations regarding them are inadequate, therefore, more reasonable invastigation is required to clarify the status of these inadequately known species.

References

Al-Saeed W.M., (2003). Pathogenic effect of *Entamoeba gingivalis* on gingival tissue of rats. *Al-Rafidain. Dent. J.,* 3: 70-73.

Arias, D.G., Regner, E.L., Iglesias, A.A., Guerrero, S.A. (2012) *Biochim. Biophys. Acta Gen. Subj.* 1820: 1859.

Ashraf et al. (2012). Prevalence of genital tract infection with *Entamoeba* gingivalis among copper T 380A instrauterine device users in Egypt. *Contraception*. 85: 108-112.

Baldauf, S. L., (2008). An overview of the phylogeny and diversity of eukaryotes. *J. Syst. Evol.* 46: 263–273.

Chia M.Y., Jeng, C.R., Hsiao, S.H., Lee, A.H., Chen, C.Y., Pang, V.F. (2009). *Entamoeba* invadens Myositis in a Common Water Monitor Lizard (Varanus salvator). V*et Pathol* 46:673- 676.

Clark, C.G., (2000). The evolution of *Entamoeba*, a cautionary Tale. *Res. Microbiol.* 151, 599–603.

Diamond L.S, Clark, C.G. (1993)."A redescription of *Entamoeba histolytica* schaudinn, 1903 (emended Whistolytic1 911) separating it from *Entamoeba dispar* brumpt, 1925". *J E Microbiol* 40:1340–4.

Debnath A, Parsonage D, Andrade RM, He C, Cobo ER, Hirata K, et al. . A high-throughput drug screen for Entamoeba histolytica identifies a new lead and target. *Nat Med.* (2012) 18:956–60. 10.1038/nm.2758

Desowitz R.S., Barnish G. (1986). *Entamoeba polecki* and other intestinal protozoa in Papua New Guinea Highland children. *Ann Trop Med Parasitol* 80:399–402.

Dobell, C. (1919). *The Amoebae Living in Man. A Zoological Monograph.* London: J. Bale, Sons, and Danielson.

Dolabella, S.S., Serrano-Luna J., Navarro-Garcı'a F., et al. (2012). Amebic liver abscess production by *Entamoeba dispar*. *Ann Hepatol* 11:107–17.

Donaldson M., Heyneman D., Dempster R., Garcia L. (1975). Epizootic of fatal amebiasis among exhibited snakes: epidemiologic, pathologic, and chemotherapeutic considerations. *Am J Vet Res* 36:808–817.

Eichinger D., (1997). Encystation of E *Entamoeba* parasites. *Bioessays.* 19: 633-639.

Elsheikha HM., Clark, G. (2018). Novel *Entamoeba* findings in nonhuman primates. *Trends in Parasitology*. 34: 283-294.

Elsheikha H.M., Clark, G. (2018). Novel *Entamoeba* Findings in Nonhuman Primates. *Trends in Parasitology* 34: 283-294.

Espinosa-Cantellano M., Mart´ınez-Palomo, A. (2000). Pathogenesis of intestinal amebiasis: from molecules to disease. *Clinical Microbiology Reviews*, 13: 318–331.

Everard, A. Lazarevic, V. Gai¨a, N., et al. (2014). Microbiome of prebiotic- treated mice reveals novel targets involved in host response during obesity. *ISME J.* 8: 2116–30.

Elsheikha, H.M., Regan, C.S., Clark, C.G. (2018). Novel *Entamoeba* findings in nonhuman Primates. *Trends Parasitol.* 34: 283-294.

Fahey, R.C., Newton, G.L. Arrick, B. Overdank-Bogart, T., Aley, S.B. (1984) *Science* 224: 70.

Fauatt, D.M., Markiewicz J.M., Danckaert A., Soubugou G., Guillen N. (2011) Human liver sinusoidal endothelial cells respond to interaction with *Entamoeba* histolytic1 by changes in morphology, integrin signalling and cell death. *CELL Microbiol* 13: 1091-1106.

Feng, M., Cai, J., Min, X., Fu, Y., Xu, Q., Tachibana, H., Cheng, X. (2013). Prevalence and genetic diversity of *Entamoeba* species infecting macaques in southwest China. *Parasitology Res.* 112: 1529-36.

Garcia, G., Ramos, F., Perez, R.G., Yanez, J., Estrada, M.S., Mendoza, L.H., Martinez-Hernandez, F., Gaytan, P., (2014). Molecular epidemiology and genetic diversity of *Entamoeba* species in a chelonian collection. *J. Med. Microbiol.* 63, 271–283.

Geiman Q.M., Rateliffe H.L. (1936). Morphology and life cycle of an amoeba producing amoebiasis in reptiles. *Parasitology* 28: 208-228.

Haq, A. et al. (1985) Experimental infection of rhesus monkeys with *Entamoeba histolytica* mimics human infection. *Lab. Anim.Sci.* 35: 481–484

Haque R., Huston C.D., Hughes M., Houpt E., et al. (2003). Current concepts: amebiasis. N. *Engl. J.* Med 348-1565-1573.

Haque R, Neville LM, Hahn P, et al. (1995). Rapid diagnosis of *Entamoeba* infection using the *Entamoeba* and *Entamoeba histolytica* stool antigen detection kits. *J Clin Microbiol.* 33: 2558-2561.

Holmgren, A., Johansson, C, Berndt, C., Lonn, M.E., Hudemann, C., Lillig, C.H. (2005). *Biochem. Soc. Trans.* 33: 1375.

Loomis, M.R., Britt J.J., Gendron A.P., Holshush H.J., Howard E.B. (1983). Hepatic and gastric amoebiasis in black and white colobus monkeys. *J. Am Vet Med Assoc.* 183: 1188-1191.

Mac Neill, A.L., Uhl, E.W., Kolenda-Roberts, H., Jacobson, E., (2002). Mortality in a wood turtle (*Clemmys insculpta*) collection. *Vet. Clin. Path.* 31, 133–136.

McCoy, J.J., Mann, B.J., Petri, W.A. Jr. (1994). Adherence and cytotoxicity of *Entamoeba histolytica* or how lectins let parasites stick around. *Infect Immun* 62:3045–3050,

Marion, S., Guillén, N. (2006). Genomic and proteomic approaches highlight phagocytosis of living and apoptotic human cells by the parasite *Entamoeba histolytica*. *Int J Parasitol.* 36:131–139.

Martinez-Palomo A., (1987). The pathogenesis of amoebiasis, *Parasitology Today.* 3: 111–118.

McHardy I.H., Wu M., Shimizu-Cohen R., Couturier M.R., Humphries R.M. (2014). Detection of intestinal protozoa in the clinical laboratory. *J Clin Microbiol.* 52:712-20.

Neal, R.A. (1950). An experimental study of *Entamoeba* muris (Grassi, 1879); its morphology, affinities and host-parasite relationship. *Parasitology*, 1950, 40, 343-365.

Kojimoto, A., Uchida, K., Horii, Y., Okumura, S., Yamaguch R., Tateyama, S., (2001). Amebiasis in four ball pythons, *Python reginus*. *J. Vet. Med. Sci.* 63: 1365–1368.

Kofoid C.A., Hinshaw H.C., Johnsotne H.G. (1929). Animal parasites of the mouth and their relation to dental disease. *J Am Dent Assoc.*1434-55.

Komatsu, T., Marsubayashi, M., Murakoshi. N., Saksi. K., Shibahara, T. (2019). Retrospective and histopathological studies of *Entamoeba* spp. and other pathogens associated with diarrhea and wasting in pigs in aichi prefecture, Japan. *JARQ* 53: 59-67.

Hamad N.M., Elkhairi, M.E., Elfaki, T.M. (2017). *Entamoeba coli* as a Potent Phagocytic Microorganism i.e., *Global Journal of Medical Research: C Microbiology and Pathology.* 17.

Moncada, D. Keller, K. Chadee, K. (2003). *Infect. Immun.* 71: 838-844.

Ponce-Gordo, F., Martínez-Díaz, R.A., Herrera, S. (2004). *Entamoeba struthionis* n.sp. (Sarcomastigophora: Endamoebidae) from ostriches (*Struthio camelus*). *Vet. Parasitol.* 119: 327–335.

Rousset, J.J. and Lauvergeat, J.A. (1971). *Protozoaires buccaux. Press Midicale* 79: 1495-1497.

Salaki J.S., Shirey J.L., Strickland G.T. (1979). Successful treatment of symptomatic *Entamoeba polecki* infection. *Am J Trop Med Hyg* 28:190.

Silberman J.D., Clark C.G., (1999). Diamond LS, et al. Phylogeny of the genera *Entamoeba* and *Endolimax* as deduced from small subunit ribosomal RNA sequences. *Mol Biol Evol.* 1999; 16:1740-1751.

Shimokawa, C., Kabir, M., et al. (2012). *Entamoeba moshkovaskii* is associated with diarrhea in infants and causes diarrhea and colitis in mice. *J. Infect Dis.* 206: 744-51.

Shiratori, T., Ishida, K., *Entamoeba* marina n. sp., a new species of *Entamoeba* isolated from tidal flats sediment of Iriomote Island, Okinawa, Japan. (2016). *J. Eukaryot Microbiol.* 63: 280-286.

Stensvold, C.R., Winiecka-Krusnell, J., Lier, T., Lebbad, M. (2018). Evaluation of a PCR Method for Detection of *Entamoeba polecki*, with an Overview of Its Molecular Epidemiology. *Journal of Clinical Microbiology.* 56: e00154-18.

Stanley Jr., S.L. (2003). Amoebiasis *Lancet*, 361: 1025–1034.

Symeonidoua, I., Diakoua, A. Papadopoulosa, E., Ponce-Gordob, F. (2004). Endoparasitism of Greek ostriches: First report of *Entamoeba struthionis* and *Balantioides coli. Veterinary Parasitology* 119: 327–335.

Tachibana H., Yanagi H., Pandey K., Cheng, X.J., Kobayashi S., Sherchand J.B., Kanbara, H. (2007). An *Entamoeba* sp. Stain isolated from rhesus monkey is virulantbut genetically different from *Entamoeba histolytica*. *Mol. Biochem. Parasitol.* 157: 7-14.

Welter B.H., Sehron MG. Temesvati L.A., et al. (2017). Flow cytometric characterization of encystation in *Entamoeba* invadens. *Mol Biochem Parasitol*. 2018: 23-27.

Wiwanitkit, Viroj (2004). *Entamoeba polecki* as a cause of human infection. *Reviews in Medical Microbiology*. 15: 41-43.

In: *Entamoeba*
Editor: Thomas L. Johnson
ISBN: 978-1-53618-506-5

Chapter 3

INFECTION BY *ENTAMOEBA HISTOLYTICA* AND IMMUNITY AGAINST THIS ENTERIC PARASITE

***Eileen Uribe-Querol*[1] *and Carlos Rosales*[2,*]**
[1]División de Estudios de Posgrado e Investigación, Facultad de Odontología, Universidad Nacional Autónoma de México, Mexico City, Mexico
[2]Departamento de Inmunología, Instituto de Investigaciones Biomédicas, Universidad Nacional Autónoma de México, Mexico City, Mexico

ABSTRACT

Entamoeba histolytica is a protozoan parasite with high prevalence in developing countries that causes amoebiasis. This disease affects the intestine and the liver; and it is the third leading cause of human deaths among parasite infections. Establishing an amoebic infection involves several crucial steps, such as degradation of the mucosal layer, adherence to the intestinal epithelium, invasion into the tissues, and dissemination to

* Corresponding Author's E-mail: carosal@unam.mx.

other organs. Each one of these infection steps depends on pathogenic factors that allow invasion and destruction of the host tissues. For invasion, the main pathogenic factors have been described and include galactose (Gal) and N-acetyl-D-galactosamine (GalNAc) lectins, glycosidases, proteolytic enzymes, and parasite-produced cytokines. For tissue destruction the amoeba uses two main mechanisms, contact-dependent cytolysis, and trogocytosis, in which amoebas nibble pieces of live cells. Once amoebas invade the tissues, they have to survive in the host by evading and confronting the immune system. Antibody production is induced by amoebas, but immunoglobulins (Ig) G do not confer protection. IgA antibodies are a major component of the human intestinal defense mechanism, and they have been shown to confer some protection in animal models of amoebiasis. Also, interferon-gamma (IFN-γ) production is involved in clearance of infection, probably via activation of leukocytes. In addition, *E. histolytica* infection of the intestine or liver is associated with a strong inflammation characterized by a large number of infiltrating neutrophils. Consequently, it has been suggested that neutrophils play a protective role in amoebiasis. However, some reports indicate that trophozoites making direct contact with neutrophils provoke lysis of these leukocytes, resulting in the release of their lytic enzymes, which in turn provoke tissue damage. Therefore, the role of neutrophils in this parasitic infection remains controversial. Recently, it was found that *E. histolytica* trophozoites are capable of inducing neutrophil extracellular traps (NET) formation, while the non-pathogenic *Entamoeba dispar*, which cannot be distinguished from *E. histolytica* by microscopic analysis, was not able to induce NET formation. Thus, NET formation seems to be a novel defense mechanism against amoebiasis. In this chapter, we will describe, the general pathogenic factors of *E. histolytica,* and how these factors participate in establishing infection. Also, we will summarize the latest knowledge on immune response and immune evasion during amebiasis.

Keywords: *Entamoeba histolytica*, amoebiasis, inflammation, immunity, neutrophil, NETosis

INTRODUCTION

Amoebiasis is a disease of the intestinal tract caused by infection with the protozoan parasite *Entamoeba histolytica*. This disease affects approximately 50 million people worldwide (Shirley et al., 2018), and according to the 2013 Global Burden of Disease Study it represents the third leading cause of death among intestinal infections (Herricks et al., 2017; Shirley et al., 2018). Amoebiasis is a serious health problem in tropical zones of the world, particularly in developing countries, where *E. histolytica* is endemic (Marie and Petri, 2014). Infections with this parasite are also found in developed countries principally among high-risk groups, such as immigrants and travelers returning from countries in endemic zones (Ross et al., 2013; Duplessis et al., 2017), men who have sex with men (Zhou et al., 2013; Shirley et al., 2018), and patients with HIV (Hung et al., 2008, Watanabe et al., 2011). The symptoms of amoebiasis in humans include diarrhea, dysentery, and colitis. However, in some cases, virulent forms of *Entamoeba histolytica* can move out of the intestine and colonize other organs, including liver, lung, and even brain. In these organs, amoebas form amoebic abscesses.

Amoebas infect people through contaminated water and food. When they reach the intestinal track, amoebas live in the outer mucus layer feeding on bacteria and many times causing only a mild disease. Sometimes however, more pathogenic amoebas can invade other tissues in conditions that are not completely recognized. The progress of an amoebic infection depends on a subtle balance between pathogenicity factors of amoebas and the immune response of the infected person. In this chapter, we summarize the current knowledge on the pathogenicity of *E. histolytica,* and on the immune response during amoebiasis.

ENTAMOEBA HISTOLYTICA LIFE CYCLE

Entamoeba histolytica has is minimal life cycle that involves only two phases, the dormant cyst and the vegetative trophozoite (Figure 1). The life cycle begins with the ingestion of mature *E. histolytica* cysts present in fecal-contaminated food or water (Mortimer and Chadee, 2010; Cornick and Chadee, 2017). In the small intestine lumen, cysts release active and motile trophozoites (Ximénez et al., 2011). Trophozoites then travel to the large intestine, where they adhere to the mucus layer and feed on bacteria of the intestinal microbiome. Trophozoites (amoebas) grow and divide through binary fission in the intestinal colon, where they can reach high densities. In these conditions, some trophozoites aggregate and activate the process of encystation (Krishnan and Ghosh, 2018). The two forms of amoebas are eliminated in feces, but only cysts can continue the life cycle of the parasite thanks to their protecting cell wall (Haque et al., 2003) (Figure 1). Thus, the life cycle of *E. histolytica* and the transmission of amoebiasis are completed through fecal-oral dissemination (Mortimer and Chadee, 2010; Marie and Petri, 2014). Because of the simplicity of its life cycle, *E. histolytica* is classified by the Centers for Disease Control and Prevention as a Class B agent. This means that *E. histolytica* is a pathogen for which monitoring and diagnosis should be improved, since it can be transmitted easily and has a major impact on morbidity and mortality (United States Centers for Disease Control and Prevention, 2019).

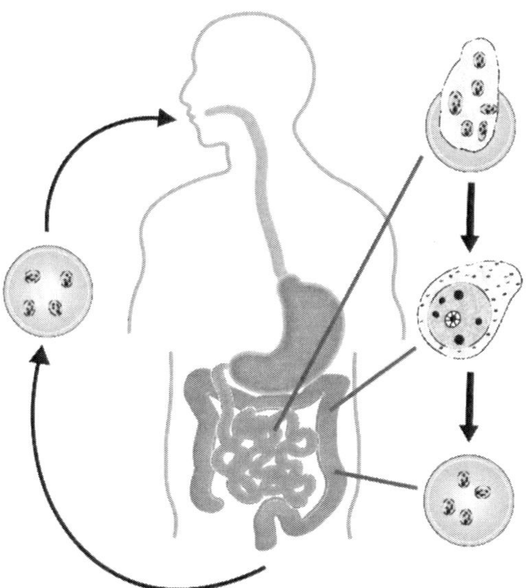

Figure 1. *Life cycle of Entamoeba histolytica*. Transmission of amoebiasis begins with the ingestion of *E. histolytica* cysts found in fecal-contaminated food or water. In the small intestine trophozoites are released and then they migrate to the large intestine, where trophozoites feed on bacteria. Some, trophozoites form new cysts, which are shed in feces to continue the transmission of the parasite.

As indicated above, most amoebic infections (around 90% of cases) cause very little damage and are self-limiting (Haque et al., 2003). However, in some infected individuals, under conditions that are not yet known, trophozoites in the intestinal colon disturb the intestinal barrier and invade other tissues. Destruction of the intestinal barrier and blood vessels causes loss of water (diarrhea) and presence of blood in the stools. Therefore, diarrhea with blood in the stools is a characteristic of severe amoebiasis. Trophozoites can then travel via the portal vein and reach distant organs including liver, lung, or brain (Dolabella et al., 2012; Wuerz et al., 2012) (Figure 2). Liver infection, causing amoebic liver abscesses is the most common manifestation of extra-intestinal amoebiasis (Wuerz et al., 2012; Prakash et al., 2020 Jan-).

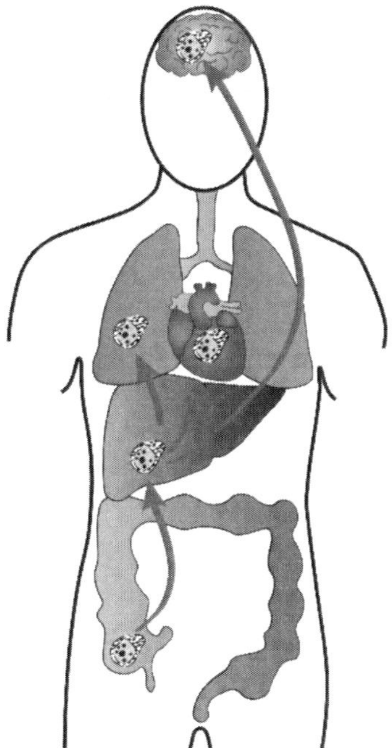

Figure 2. *Entamoeba histolytica can invade multiple tissues.* Trophozoites residing in the colon disrupt the intestinal barrier and subsequently disseminate via the afferent blood flow of the portal vein to reach distant sites such as liver, lung, or brain. In particular, amoebic liver abscesses are the most common manifestations of extra-intestinal amoebiasis.

CLINICAL MANIFESTATIONS

Luminal Amoebiasis

The greater part of individuals infected with *Entamoeba histolytica* are asymptomatic. Only 10% to 20% of infected individuals present symptoms of disease. The reasons for this are not understood, but several risk factors associated with disease severity have been identified. They include lack of urban services, young age, malnutrition, alcoholism, corticosteroid use, cancer, (Haque et al., 2003; Wuerz et al., 2012; Singh et al., 2019), and even the host gut microbiome (as discussed latter) (Ngobeni et al., 2017b; Singh et al., 2019). When symptoms appear during intestinal amoebiasis, they vary from mild diarrhea to severe dysentery.

The fact that very few infected individuals develop amoebiasis strongly suggests that genomic variability exists among amoeba isolates and also among humans. Several genetic studies of DNA nucleotide polymorphisms have confirmed that substantial genomic variability exists among *E. histolytica* strains with different geographic mobility (Haghighi et al., 2002; Ramos et al., 2005), and also between pathogenic and non-pathogenic cultures of *E. histolytica* (Balderas-Renteria et al., 2007). These DNA polymorphisms are potential candidates to become genetic markers of pathogenicity for different amoeba isolates. In the case of humans, few genetic associations with amoebiasis have been reported. Association between antigens of the major histocompatibility complex (MHC) particularly class II antigens (HLA) and amoebiasis was suggested in a study of Bangladeshi children (Duggal et al., 2004) and in a study of two Mexican populations from different geographical areas (Hernández et al., 2015). In the Bangladeshi children, a potential protective association was observed with the HLA allele DQB1*0601 (Duggal et al., 2004). In the Mexican study, the HLA-DQB1*02 allele was present with higher frequency in patients with amoebic liver abscesses than in healthy individuals (Hernández et al., 2015). A stronger association between predisposition to *E. histolytica* infection and a genetic marker was found in the leptin receptor gene. Leptin is a hormone produced by fat-tissue cells that inhibits eating. Because Leptin production is reduced in malnourished children, which also present increased susceptibility to *E. histolytica* infections, an association between leptin and amoebiasis was investigated. A mutation in the leptin receptor gene was found to be associated with a higher predisposition to *E. histolytica* infections (Duggal et al., 2011). Children carrying the allele for arginine (223R) were nearly four times more likely to have an amoeba infection than those homozygous for the more common glutamine allele (223Q) (Duggal et al., 2011). In addition, the association between this polymorphism and amoebiasis was also confirmed in animal models. Mice deficient in the leptin receptor became very sick with complete intestinal mucosal destruction after *E. histolytica* infection (Guo et al., 2011b). Moreover, 40% of mice carrying the humanized leptin receptor arginine (223R) allele were infected, whereas all mice carrying the glutamine (223Q) allele eliminated the amoebas without signs of disease (Mackey-Lawrence et al., 2013).

Disseminated Amoebic Disease

In some cases, trophozoites can damage the intestinal barrier and migrate to other organs causing extra-intestinal amoebiasis (Figure 2). This form of amoebiasis is most common in the liver, where trophozoites form amoebic liver abscesses (Prakash et al., 2020 Jan-). Patients with this form of the disease present fever and continuous right upper quadrant pain (Wuerz et al., 2012). Also, about half of these patients develop protracted diarrhea, dysentery, weight loss, and abdominal pain (Mortimer and Chadee, 2010). An intriguing characteristic of amoebic liver abscesses is that they appear in young adults (20 to 40 years of age) and are 10 times more frequent in men than in women (Acuna-Soto et al., 2000). The reasons for this gender difference are not clear. One possible explanation is a more efficient complement activation in women. *E. histolytica* killing was mediated by complement (Snow et al., 2008), and the serum from women was significantly more effective at killing trophozoites than serum from men (Snow et al., 2008). Another possible explanation for this gender bias is that testosterone levels seem to increase the susceptibility to amoebic liver abscess in mice through inhibition of interferon-gamma (IFN-γ) secretion (Lotter et al., 2013). After rupture of a liver abscess, amoeba can also reach the lungs. These organs are the second most frequent site invaded by amoebas (Shamsuzzaman and Hashiguchi, 2002) (Figure 2). From lungs, amoebas can also reach the heart, causing a rare complication with a high mortality rate (Nunes et al., 2017). Another infrequent form of extra-intestinal amoebiasis is the formation of a brain abscess. This form of the disease is usually lethal (Maldonado-Barrera et al., 2012).

Laboratory Diagnosis

Intestinal diseases presenting diarrhea are still an important factor for high morbidity and mortality among children in developing countries (Herricks et al., 2017). *E. histolytica* is a parasite that significantly contributes to these diseases. Therefore, it is important to improve the diagnosis for this parasite (United States Centers for Disease Control and Prevention, 2019). Several diagnostic methods have been developed for revealing the presence of *E. histolytic.* These methods include microscopy, serology (Fotedar et al., 2007), antigen detection (Haque et al., 1993), nucleic acid detection by polymerase chain reaction (PCR) (López-López et al., 2017; Ryan et al., 2017), and colonoscopy (Table 1). Microscopy is the simplest method and the most commonly used in many places. However, microscopy is not recommended nowadays as a reliable diagnostic tool for amoebiasis. The recommended approach for diagnosing *E. histolytica* is to perform a combination of assays in order to improve the specificity and sensitivity of the diagnosis for amoebiasis (Ryan et al., 2017). Antibody detection, antigen detection, and PCR are the most reliable methods (Ryan et al., 2017). Unfortunately, using a combination of assays is not possible in many endemic areas due to limited resources.

Table 1. Diagnostic methods for *E. histolytica*

Method	Advantages	Disadvantages	Sensitivity (%)	Specificity (%)
Microscopy	Widely available	Low sensitivity and specificity	< 60	--
	Low cost	Several stool samples required		
	Minimal equipment	Skilled observer required		
	Detection of other parasites	time-consuming		
Serology	High sensitivity and specificity	Antibodies remain in serum for years after resolution of infection. Thus, not very helpful in endemic areas; more useful in diagnosing travelers	65 – 90	> 90
Antigen detection	Simple and rapid to perform, commercially available	Requires fresh, not fixed stool samples for analysis	0 – 88	> 80
	High sensitivity in endemic areas but reduced sensitivity in nonendemic areas	Poor sensitivity for amebic liver abscess		
Polymerase Chain Reaction (PCR)	High sensitivity and specificity both for intestinal and extra intestinal infections (Gold standard)	More expensive Limited use in poor endemic areas Requires special instruments, kits, and skilled technicians	92 - 100	89 - 100
Colonoscopy	Can detect intestinal lesions directly	Expensive. Requires trained physicians. Limited use in poor endemic areas	50 - 80	> 60

PATHOGENIC MECHANISMS OF *ENTAMOEBA HISTOLYTICA*

E. histolytica cysts are ingested with contaminated food. Cysts are resistant to acid in the stomach so they can safely arrive at the small intestine. In there, cysts release trophozoites which then travel the large intestine and adhere to the mucus layer of the colon. Trophozoites feed on commensal bacteria and can live there without inducing symptoms of disease. However, in certain conditions, trophozoites can destroy the mucus barrier and adhere directly to the surface of intestinal epithelial cells. The conditions that favor a disequilibrium between the microbiome and amoebas are still unknown. A change in the type of bacteria (dysbiosis) is probably a major factor (see later). After adhesion, trophozoites are able to destroy epithelial cells and cause inflammation. Amoeba phagocytosis and trogocytosis are important pathogenic factors during this process. Next, amoebas can degrade the extracellular matrix and disrupt epithelial tight junctions, which allow amoebas to travel into the extra-intestinal space and, sometimes, disseminate to other organs. Tissue destruction and inflammation are important signals for recruitment of immune cells, mainly neutrophils.

Microbiome and Dysbiosis

In most cases of *E. histolytica* infection, amoebas colonize the outer mucus layer of the intestine away from epithelial cells. In that area, amoebas feed on bacteria and sugars from mucus and create an infection with no symptoms of disease (Figure 3). In this situation, the mucus layer plays a protective role in maintaining a healthy equilibrium between the microbiome and pathogens (Pelaseyed et al., 2014; Sperandio et al., 2015). Adhesion of *E. histolytica* to the mucus layer of the intestine, is mediated by a surface lectin (Stanley, 2003) with high affinity for galactose (Gal) and N-acetyl-D-galactosamine (GalNAc) oligosaccharides on MUC2 mucin, the major component of the mucus layer (Birchenough et al., 2015). Intestinal bacteria are important for amoebas, not only as a source of nutrients but also as a contributing factor of pathogenicity (Burgess and Petri, 2016). Infection of germ-free animals with *E. histolytica* did not cause disease. However, in the presence of bacteria amoebas could induce disease (Phillips et al., 1955). Similarly, *E. histolytica* trophozoites from axenic cultures were less infective than trophozoites cultured in the presence of live bacteria (Wittner and Rosenbaum, 1970). Amoebas feed on bacteria through phagocytosis, and they preferentially ingest *Lactobacillus* commensal bacteria (Iyer et al., 2019).

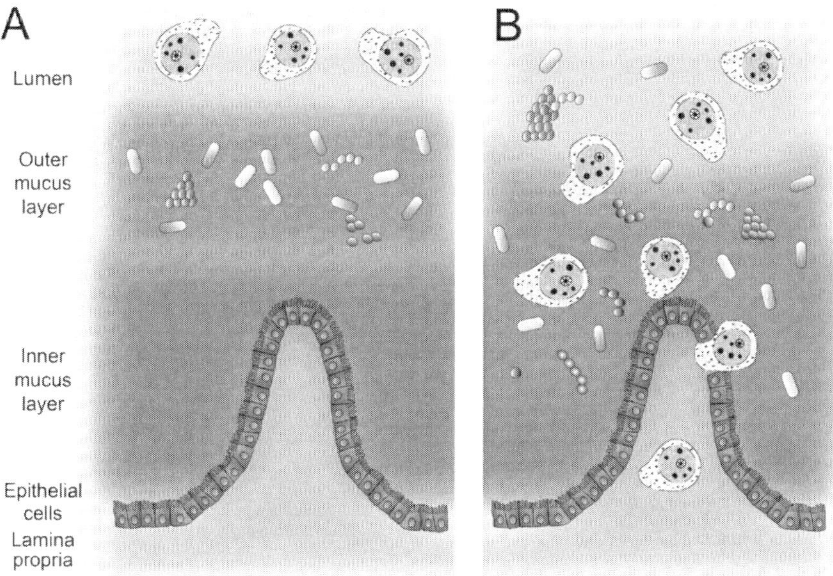

Figure 3. *Entamoeba histolytica disrupts the mucus layer and cause invasive disease.* (A) When *E. histolytica* infects a healthy individual with an intact intestinal barrier, amoebas settle in the outer mucus layer away from the epithelium. Here, amoebas feed on bacteria and sugars from mucus and establish an asymptomatic infection. Thus, mucus and bacteria play a protective role by maintaining a healthy equilibrium between the microbiome and amoebas. (B) Under undetermined conditions, *E. histolytica* breaks innate host defenses by degrading the mucin layer with various glycosidases and cysteine proteinases. *E. histolytica*-induced mucus degradation exposes intestinal epithelial cells and alters the microbiome composition (dysbiosis). Amoebas then bind to epithelial cells, inducing mucus and water secretion and stimulating an acute inflammatory response. Next, epithelial cells die and amoebas can initiate the active process of invasion into tissues.

However, when a healthy microbiome is altered (dysbiosis) other bacteria appear to be related to pathogenesis of *E. histolytica* (Guillén, 2019; Leon-Coria et al., 2020). A dysbiotic state, with an increase in *Bifidobacterium* species and a decrease in *Bacteroides*, *Lactobacillus*, *Campylobacter* and *Eubacterium* species was associated with disease in *E. histolytica*-infected patients from India (Verma et al., 2012). Similarly, an increase in the dysbiotic bacteria *Prevotella copri* was associated with amoeba-induced diarrhea of children from Bangladesh (Gilchrist et al., 2016). Also, anaerobic bacteria species such as *Fusobacterium, Peptococcus, Prevotella*, and *Bacteroides* were associated with amoebic liver abscess in patients from India (Singh et al., 2019). Thus, dysbiosis is clearly a risk factor for amoebiasis (Leon-Coria et al., 2018; Leon-Coria et al., 2020) (Figure 3).

Amoeba Adhesion to Intestinal Epithelium

When dysbiosis exists, amoebas can degrade the mucin layer by the action of glycosidases and proteinases (Moncada et al., 2005; Lidell et al., 2006). The mucin carbohydrate side chains are digested by various *E. histolytica*-secreted glycosidases, such as N-acetylgalactosamidase, N-acetyl-glucosaminidase, β-galactosidase, and β-N-acetyl-hexosaminidase (Frederick and Petri, 2005; Thibeaux et al., 2013). Then, the protein backbone of mucin can be degraded by cysteine proteinases. *E. histolytica* express mainly four cysteine proteinases, EhCP-A1, EhCP-A2, EhCP-A5, and EhCP-A7 in cultures (Irmer et al., 2009). Among these proteinases, EhCP-A5 plays a central role in pathogenicity (Tillack et al., 2006; Freitas et al., 2009). Also, it was recently reported that EhCP-A5 also activates host matrix metalloproteinases (MMP), which efficiently degrade extracellular matrix proteins (Thibeaux et al., 2014). In addition, EhCP-A5 contains an arginine-glycine-aspartate (RGD) motif outside its catalytic site that becomes an integrin-binding site. Binding of an EhCP-A5 RGD peptide to αvβ3 integrin on goblet cells, induces water and mucus secretion that leads to goblet cells cavitation and mucus depletion (Cornick et al., 2016). Finally, destruction of the mucus layer leaves the intestinal epithelial cells exposed. *E. histolytica* can then adhere to unprotected cells via its Gal/GalNAc lectin which binds to Gal and GalNAc residues on the membrane of epithelial cells.

Epithelial Cell Lysis

When amoeba trophozoites adhere to epithelial cells, amoebas are capable of causing host cell destruction by a direct cell-contact mechanism or by releasing lytic molecules. The Gal/GalNAC lectin is the main adhesion molecule used by *E. histolytica* (Ankri et al., 1999). However, other molecules such as a 220 kDa lectin, a 112 kDa adhesin/cysteine protease (Rodríguez et al., 1989), and a family of serine-rich proteins (SREHPs) with tandem repeats are also involved in adhesion to host cells (Aguirre García et al., 2015). The Gal/GalNAC lectin is important for the contact-dependent cytotoxic activity of *E. histolytica*, since inhibiting this lectin expression blocked the cytotoxic activity (Ankri et al., 1999). The mechanism by which amoeba adhesion to cells induce cell lysis is not completely elucidated. However, it seems that after adhesion, cytotoxicity is mediated mainly by apoptosis (Ghosh et al., 2019). Even though the signaling pathway that initiates apoptosis in host cells is not completely described (Huston et al., 2000), it seems to involve elevation of intracellular calcium leading to activation of caspase-3 (Ralston and Petri, 2011). Also, amoeba-induced activation of K^+ channels seems to be involved, since adhesion of trophozoites provoke an efflux of K^+ ions leading to cell lysis (Marie et al., 2015).

In addition, after cell adhesion, trophozoites can release molecules with cytotoxic effect, such as amoebapores, cysteine proteinases, and phospholipase A2. Amoebapores are soluble proteins that belong to the functionally diverse family of saponin-like proteins (SAPLIPs), which have the capacity to interact with lipids (Bruhn and Leippe, 2001) and form oligomeric pores in target cell membranes (Andrä et al., 2003). The main function of amoebapores is lysis of phagocytosed bacteria

(Winkelmann et al., 2006). However, when secreted amoebapores can also induce lysis of epithelial cells and leukocytes (García-Zepeda et al., 2007). Also, the cysteine proteinases participate in degradation of the basement membrane, facilitating amoeba invasion into tissues (Nakada-Tsukui and Nozaki, 2016), and phospholipase A2 participates in lysis of erythrocytes (Vargas-Villarreal et al., 1998).

Phagocytosis and Trogocytosis

E. histolytica displays very efficient phagocytosis, and this processes is also an important pathogenic factor (Labruyère and Guillén, 2006; Christy and Petri, 2011; Labruyère et al., 2019). Phagocytosis of erythrocytes (Trissl et al., 1978), bacteria (Orozco et al., 1983), and dead cells (Huston et al., 2003) is vital for these amoebas. *E. histolytica* isolates having a higher capacity to phagocytose erythrocytes also presented higher pathogenic potential (Trissl et al., 1978). Similarly, *E. histolytica* clones having phagocytosis defects were also less pathogenic (Orozco et al., 1983; Rodríguez and Orozco, 1986). Amoebas prefer to phagocytose apoptotic cells, implying that trophozoites induce apoptosis of cells before phagocytosing them (Huston et al., 2003). Yet, trophozoites in amoebic liver abscess in hamsters were shown to ingest hepatocytes (Blazquez et al., 2007). Phagocytosis receptors on amoebas remain unidentified. However, some membrane molecules are proposed to be phagocytic receptors. For example, the Gal/GalNAc lectin has been reported to mediate phagocytosis of erythrocytes (Katz et al., 2002). Other candidates include the classical G-protein-coupled receptor (GPCR), EhGPCR-1 which concentrates in phagocytic cups during erythrocyte phagocytosis (Picazarri et al., 2005), and the EhROM1 which is found together with the Gal/GalNAc lectin during phagocytosis (Baxt et al., 2008; Baxt et al., 2010). Also, the serine-rich *E. histolytica* protein (SREHP) has been reported to mediate binding to apoptotic lymphocytes (Teixeira and Huston, 2008).

Phagocytosis is a fundamental biological process (Rosales and Uribe-Querol, 2017; Uribe-Querol and Rosales, 2020) and as a result there is much interest in revealing the signaling pathways controlling this process in amoebas. In the past few years, important progress has been made in this topic. The Rho GTPases are known regulators of the actin cytoskeleton in mammalian cells (Mao and Finnemann, 2015). Similarly, the *E. histolytica* EhRho1 was found to regulate phagocytosis via the actin nucleating proteins EhFormin1 and EhProfilin1 (Bharadwaj et al., 2018). Also, the Eh14-3-3 Protein 3 (EhP3) was found to cluster at phagocytic cups (Agarwal et al., 2019), while the Arp2/3 protein complex was important for formation of adhesion plaques (Manich et al., 2018). These studies point to actin dynamics as an important element for cell motility and phagocytosis. Because *E. histolytica* has a single actin molecule (Manich et al., 2018), actin dynamics in amoeba is mostly regulated by post-translational modifications of the actin-rich cytoskeleton. Acetylation, followed by phosphorylation were the most important modifications of actin during phagocytosis (Hernández-Cuevas et al., 2019). Once a phagosome is formed, it changes its membrane composition and contents to become a phagolysosome in a process known as phagosome maturation (Levin et al., 2016; Pauwels et al., 2017). Rab proteins are important functional molecules that participate in directing the fusion of phagosomes with lysosomes (Gutierrez, 2013). In the case of amoebas, the amoebic EhRab21 localizes to lysosomes and also to Golgi-positive structures during phagocytosis of erythrocytes, suggesting that EhRab21 participates in the traffic between lysosomes and the Golgi apparatus (Constantino-Jonapa et al., 2018). Similarly, the EhRabB was shown to mobilize the cysteine proteinase-adhesin complex of this parasite (EhCPADH) through the actin cytoskeleton, indicating that these molecules are important for phagocytosis (Javier-Reyna et al., 2019). During phagocytosis by mammalian cells, phosphoinositides also play an essential function regulating membrane and actin remodeling (Rosales and Uribe-Querol, 2017). In the same way, the *E. histolytica* enzyme EhPIPKI, which generates PtdIns(4,5)P_2, concentrates in phagocytic cups, but is absent in closed phagosomes (Sharma et al., 2019). Similarly, the amoebic endosomal sorting complex required for transport (ESCRT)-III participates in phagocytosis by interacting with vacuolar protein sorting (Vps) proteins to assemble phagosomes (Avalos-Padilla et al., 2018). These studies help us to elucidate how pathogenesis of *E. histolytica* is regulated by phagocytosis.

Another important pathogenic factor of *E. histolytica* is trogocytosis. This is a recently discovered process by which amoebas nibble pieces of live cells (Ralston et al., 2014). The damage caused by trogocytosis finally leads to endothelial cell lysis. During trogocytosis, acidified lysosomes (Gilmartin et al., 2017) and the AGC family kinase 1 (Somlata et al., 2017) participate. Interestingly, AGC family kinase 1 is involved in trogocytosis of live human cells, but it does not participate in phagocytosis of dead cells, suggesting that the trogocytosis signaling pathway is likely to be different from phagocytosis (Gilmartin et al., 2017).

Opening of Tight Junctions and Tissue Invasion

Amoebas adhered to intestinal epithelium cells not only destroy the cells but can also induce the opening of tight junctions and the degradation of extracellular matrix proteins. Together these events allow amoebas to move into the submucosal region and invade tissues (Figure 3). Amoebas release prostaglandin E2 (PGE2), which binds to the prostaglandin E receptor-4 on enterocytes leading to disruption of tight junctions (Lejeune et al., 2011), and also to production of interleukin (IL)-8 (Dey and Chadee, 2008). Likewise, the amoebic 112 kDa adhesin can move into the intercellular space of epithelial cells and bind to the tight junctions proteins claudin-1 and occludin, resulting in opening of tight junctions (Cuellar et al., 2017). In addition, amoebas can degrade extracellular matrix proteins by releasing cysteine proteinases either directly or indirectly by activation of other proteinases. For example, EhCP-A5 can activate host matrix metalloproteinases (MMP), which efficiently degrade extracellular matrix proteins (Thibeaux et al., 2014).

IMMUNE RESPONSE TO *ENTAMOEBA HISTOLYTICA*

Once amoebas adhere to intestinal epithelial cell and begin invasion of tissues, the immune system responds with different mechanisms to protect the host from the parasite (Figure 4). In turn, amoebas display several evasion strategies to resist the immune response and continue their survival. Although *E. histolytica* infection has been known and studied for many years, our understanding of the immune response against this parasite is still very limited (Kantor et al., 2018). The principal immune mechanisms against an *E. histolytica* infection are described next (Figure 4).

Innate Immune Mechanisms

The initial defense mechanisms encountered by invading amoebas are elements of the innate immune system. Acid in the stomach is a potent microbicidal element, but amoeba cysts are able to resist low pH (Moonah et al., 2013). Once in the intestine, trophozoites encounter a layer of mucin mucus that protects epithelial cells. Amoebas bind to mucin via their Gal/GalNAc lectin and feed on commensal bacteria. However, as indicated earlier, in dysbiosis conditions, amoebas release glycosidases and cysteine proteinases that disrupt the mucus layer (Lidell et al., 2006). Then, amoebas can reach the epithelium. Intestinal epithelial cells in turn, respond by detecting the carbohydrate recognition domain of the Gal/GalNAc lectin via toll-like receptor (TLR)-2 and TLR-4 (Kammanadiminti et al., 2004; Campos-Rodríguez and Jarillo-Luna, 2005; Galván-Moroyoqui et al., 2011). Another amoeba surface molecule, the lipopeptidophosphoglycan (LPPG) (Guha-Niyogi et al., 2001) is also recognized by TLR-2, (Maldonado-Bernal et al., 2005). Detection of amoebas by these TLRs induces activation of NF-κB in the epithelial cells, which then induces the production of inflammatory cytokines, including IL-1β, IL-6, IL-8, IL-10, IL-12, IFN-γ, and tumor necrosis factor-alpha (TNF-α) (Sharma et al., 2008). All these cytokines induce inflammation and further stimulate other cells of the immune system. In particular, TNF-α has been reported to correlate with diarrhea in children infected by *E.*

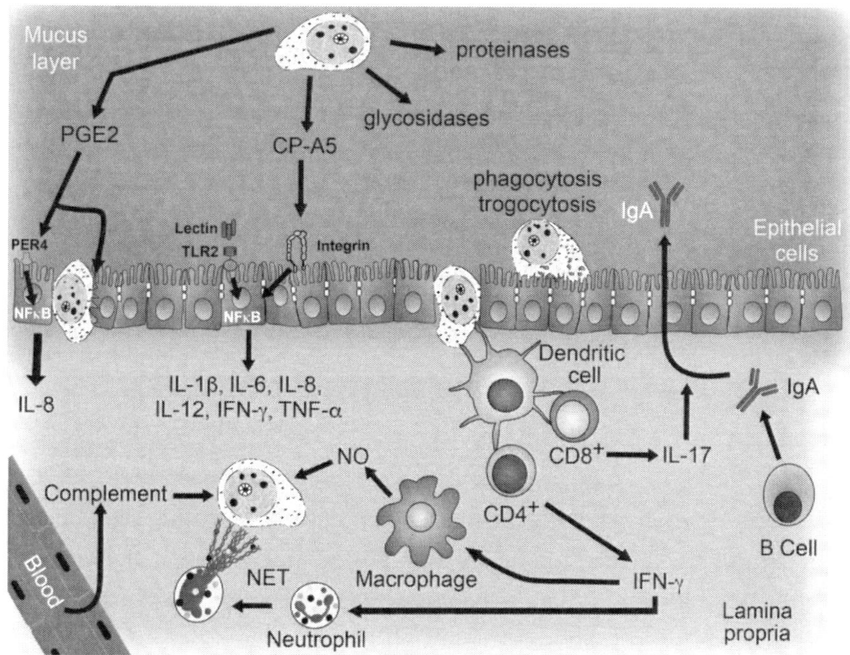

Figure 4. *Immune responses against Entamoeba histolytica*. In the lumen of the large intestine, glycosidases and proteinases secreted from amoebas are involved in the degradation of mucin. The proteinase EhCP-A5 binds to and activates integrins on endothelial cells, leading to NF-κB activation. Also, the Gal/GalNAc lectin (lectin) on the amoeba binds to toll-like receptor 2 (TLR2) leading to NF-κB activation and release of inflammatory cytokines, including interleukin 1 beta (IL-1β), IL-6, IL-8, L-12, interferon-gamma (IFN-γ) and tumor necrosis factor alpha (TNF-α). Prostaglandin E2 (PGE2) secreted from the amoeba, disrupts tight junctions of endothelial cells. PGE2 also binds to the prostaglandin E receptor-4 (PER4), activates NF-κB and stimulates IL-8 secretion. Phagocytosis and trogocytosis are also involved in removal of epithelial cells and invasion into the tissue. Infiltrating trophozoites are attacked by complement from the circulation. Amoebas are recognized by dendritic cells, which then activate $CD4^+$ and $CD8^+$ T cells, for developing a cellular immune response. $CD4^+$ T cells produce IFN-γ, while $CD8^+$ T cells produce IL-17. IFN-γ stimulates macrophages to produce nitric oxide (NO), and neutrophils to release extracellular traps (NET). NO can directly kill amoebas, while NET can trap and immobilize them. IL-17 enhances secretion of IgA antibodies (humoral immune response) into the colonic lumen.

histolytica (Peterson et al., 2010), and with severity of amoebic liver abscess in a mouse model of amoebiasis (Helk et al., 2013). Amoebas can also stimulate inflammation without even touching the epithelial cells thanks to secreted proteins that induce epithelial cells to produce cytokines (Yu and Chadee, 1997). *E. histolytica* makes a homolog of the cytokine macrophage migration inhibitory factor (EhMIF), which seems to be required for intestinal inflammation (Ngobeni et al., 2017a). In addition, Also, the PGE2 secreted by *E. histolytica* not only induces opening of cells tight junctions (Lejeune et al., 2011) but also stimulates secretion of IL-8 (Dey and Chadee, 2008). IL-8 is a chemoattractant for neutrophils, thus these leukocytes are the first immune cells to arrive at the intestinal tract during an amoebic infection (Moonah et al., 2013). In addition, amoebic cysteine proteinases display IL-1β converting enzyme activity, that leads to IL-1β production (Zhang et al., 2000). Together, these reports indicate that inflammation helps amoebas to establish an infection and to invade tissues (Mortimer and Chadee, 2010; Marie and Petri, 2014).

After breaching the intestinal epithelium, amoebas go into the circulation, where they encounter the complement system. Complement has been suggested to be relevant against amoebas by the fact that serum from women is more efficient in complement-mediated lysis of trophozoites than serum from men (Snow et al., 2008). However, amoebas are also able to resist complement in various ways (Braga et al., 1992). One way is the thick layer of LPPG within the amoeba glycocalyx that prevents the formation of the complement membrane attack complex (MAC) (Bhattacharya et al., 2000; Weber et al., 2008). Another way is the presence of a CD59-like region within the Gal/GalNAc lectin. Similarly to CD-59 the amoeba lectin inhibits the formation of a MAC on the amoeba (Weber et al., 2008). Also, it was recently found that amoebas can acquire host cell membrane proteins through trogocytosis and expose them on its membrane. This process then results in amoebas becoming resistant to complement-mediated lysis (Miller et al., 2019). Yet, another way is the fact that amoeba cysteine proteinases can also degrade complement components (Reed et al., 1995).

The strong inflammatory reaction induced by *E. histolytica* infection is also characterized by the presence of many neutrophils (Espinosa-Cantellano and Martínez-Palomo, 2000; Campos-Rodríguez et al., 2016). Neutrophils, the most abundant leukocytes in circulation, migrate from blood to sites of infection, where they exhibit antimicrobial functions (Mayadas et al., 2014; Rosales, 2020), including phagocytosis (Rosales and Uribe-Querol, 2017; Uribe-Querol and Rosales, 2020), degranulation, formation of reactive oxygen species (ROS) (Rosales, 2020), and formation of neutrophil extracellular traps (NET) (Neeli and Radic, 2012; Yipp et al., 2012). Neutrophils seem to be important against this amoebas, since less neutrophils results in more severe amoebiasis (Jarillo-Luna et al., 2002; Estrada-Figueroa et al., 2011). However, some reports suggest neutrophils can have adverse effects by inducing more inflammation and tissue damage (Pérez-Tamayo et al., 2006; Olivos-García et al., 2007; Dickson-Gonzalez et al., 2009) after *E. histolytica* trophozoites kill them and neutrophil lytic products are released (Sim et al., 2007). Therefore, the role of neutrophils in amoebiasis is complex and not completely understood.

The mechanisms used by neutrophils to control an amoebic infection are still not completely understood. Initially, it was been assumed that amoebas are susceptible to ROS produced by neutrophils (Choi et al., 2005; Biller et al., 2014) because amoebas were killed in vitro by H_2O_2 in an apoptosis-like manner (Ghosh et al., 2010), and amoeba peroxiredoxin can degrade ROS released from immune cells (Choi et al., 2005; Davis et al., 2006). However, there is no direct evidence that amoebas are killed by ROS, and recently it was reported that neither pathogenic *E. histolytica* (Díaz-Godínez et al., 2018; Fonseca et al., 2018) nor non-pathogenic *Entamoeba dispar* (Davis et al., 2007; Bansal et al., 2009), could induce ROS production by neutrophils (Fonseca et al., 2019). Thus, it is possible that neutrophils attack amoebas by other mechanisms besides ROS.

Another mechanism of neutrophils against microorganisms is the formation of NET, which are fiber-like structures formed by DNA covered by proteins from neutrophil granules (Papayannopoulos and Zychlinsky, 2009; Neeli and Radic, 2012). NET are preferentially formed against large microorganisms to prevent their dissemination (Branzk et al., 2014). Recently, it was shown that *E. histolytica* trophozoites are capable of inducing NET formation (Ávila et al., 2016; Díaz-Godínez et al., 2018). Interestingly, this response was independent of ROS (Díaz-Godínez et al., 2018; Fonseca et al., 2018) and was not induced by the non-pathogenic *Entamoeba dispar* (Fonseca et al., 2019). These NET were also found not only to stop amoebas from moving but also to kill them directly (Fonseca et al., 2019). Thus, it seems that neutrophils can selectively recognize pathogenic *E. histolytica* from the non-pathogenic amoeba and control dissemination of parasites by releasing NET around them (Fonseca et al., 2019). Interestingly, no evidence of apoptosis was found in neutrophils interacting with amoebas (Díaz-Godínez et al., 2018). Thus, it seems that instead of the amoebas inducing neutrophil death, it is the neutrophil that attacks the trophozoite by undergoing NETosis.

Macrophages are another innate immune cell type involved in protection from amoebiasis. Macrophages detect amoebas via TLR-2 and TLR-4, and in consequence they get activated. Activated macrophages release inflammatory cytokines, such as IL-1, IL-6, and IL-12 (Kammanadiminti et al., 2004; Maldonado-Bernal et al., 2005), and produce large amounts of nitric oxide (NO) from L-arginine, the substrate of NO synthase (NOS-II) (Mortimer and Chadee, 2010). NO is capable of killing *E. histolytica* by inhibiting essential enzymes in the metabolism of amoebas (Ramos et al., 2007). However, amoebas also display mechanisms against the NO from macrophages. First, amoebas produce monocyte locomotion inhibitory factor, which inhibits NO production (Rico et al., 2003) and second, they also produce *E. histolytica* arginase (EhArg), which promotes hydrolysis of L-arginine to L-ornithine and urea, thus reducing the substrate for NO synthesis (Elnekave et al., 2003) (Figure 4).

Humoral Immune Response

People infected with *E. histolytica* can produce amoeba-specific antibodies. In serum, IgG antibodies are found. However, these antibodies are typically not protective, since individuals having them usually present recurrent infections (Moonah et al., 2013; Marie and Petri, 2014). In addition, the amoeba cysteine proteinases are capable of degrading IgG in a dose-dependent manner (Tran et al., 1998) resulting in poor protection from IgG antibodies. In contrast, IgA antibodies are found in the intestine. These secreted IgA antibodies are protective. For example, IgA antibodies against the Gal/GalNAc lectin were found to inhibit amoeba infection of the colon (Haque et al., 2001; Haque et al., 2002; Haque et al., 2006). Similarly, IgA antibodies against the Gal/GalNAc lectin were also associated with protection from subsequent amebic infections, in patients with amoebic liver abscess (Ravdin et al., 2003; Abd-Alla et al., 2006). Also, IgA antibodies against EhMIF prevented reinfection (Ngobeni et al., 2017a). However, amoebas can also degrade IgA by the action of their cysteine proteinases in the intestine lumen (Garcia-Nieto et al., 2008; Mortimer and Chadee, 2010).

Cellular Immune Response

Cell-mediated immune responses are also important for host defense against *E. histolytica*. Dendritic cells can detect amoebas via TLRs (Wong-Baeza et al., 2010) and then they activate T-cell lymphocytes (Sharma et al., 2008) (Figure 4). $CD4^+$ T lymphocytes were shown to contribute to establish an intestinal amoebic infection in C3H/HeJ mice. Depletion of $CD4^+$ T cells lead to lower number of amoebas (Houpt et al., 2002), indicating that T cell-mediated inflammation also favors the disease. However, T cells not only induce inflammation but also modulate the type of immune response depending on the Th1 or Th2 profile of cytokines they produce (Rafiei et al., 2009). IFN-γ, a Th1 cytokine, offers protection from amebiasis (Sánchez-Guillén et al., 2002; Haque et al., 2007; Deloer et al.,

2017; Gonzalez Rivas et al., 2018), while IL-4, a Th2 cytokine, is associated with acute phase of the infection and also with invasive amebiasis (Bansal et al., 2005; Guo et al., 2008; Bernin et al., 2014).

IFN-γ is associated with protection from amebiasis (Lotter et al., 2009). This cytokine is mainly produced by $CD4^+$ T cells but also by natural killer T cells (NKTs). The IFN-γ effect against amoebas is mediated predominantly by other cells, such as neutrophils and macrophages, which present amoebicidal activity *in vitro* after IFN-γ stimulation (Ghadirian and Denis, 1992; Lin and Chadee, 1992) (Figure 4). Similarly, $CD8^+$ cytotoxic T-cells induce damage to amoebas through production of IL-17 (Guo et al., 2009; Guo et al., 2011a). IL-17 promotes several responses of intestinal epithelium cells against amoebas. IL-17 induced secretion of mucin (Nishida et al., 2012) and antimicrobial peptides (Liang et al., 2006), and stimulation or IgA transport across the epithelium (Cao et al., 2012). Another cytokine associated with protection from amebiasis is IL-10. This cytokine is produced by $CD4^+$ T cells and it was shown to have a protective effect from intestinal amoebiasis in experiments with chimeras of susceptible (C3E) and resistant (B6) strains of mice (Hamano et al., 2006). IL-10 acts on intestinal epithelial cells to promote the natural resistance of B6 mice to *E. histolytica* infections (Hamano et al., 2006). In addition, macrophages activated through TLR2 and TLR4 with amoebic LPPG also produced IL-10 (Maldonado-Bernal et al., 2005).

Finally, a subpopulation of regulatory T cells, characterized by the expression of the chemokine receptor CCR9, was found in a model of amoeba infection. These regulatory T cells are proposed to have a role in controlling inflammation during *E. histolytica* infection (Rojas-López et al., 2012).

Conclusion

Amoebiasis remains a serious health problem in developing countries. Yet, many new cases appear around the world due to travel from endemic areas. During an *E. histolytica* infection, the immune system gets activated using both innate and adaptive immune responses. In turn, amoebas use multiple pathogenic factors as defense strategies. In most cases, the immune response seems to be effective in protecting the host, as indicated by the fact that only few infected individual (around 10%) progress to invasive amoebiasis (Kantor et al., 2018). Still, our understanding of the immune response against this parasite is limited. Studies on immunoglobulin, cytokine and cellular functions among asymptomatic carriers and patients with invasive disease, will help clarify how the balance between resistance and disease is controlled. Neutrophils are always present in amoebiasis lesions. How these leukocytes help controlling amoebiasis, is an open question. A possible mechanism is the production of NET, which can trap and kill amoebas (Ávila et al., 2016; Fonseca et al., 2019). The Gal/GalNAc lectin induces IgA-mediated protective immunity, and as such, it is considered a suitable antigen for vaccine development (Singh et al., 2016; Abhyankar et al., 2018). Finally, a combination of education campaigns for improving sanitation, and basic research on the pathogenesis of *E. histolytica* infection will improve our chances to one day, eradicate amoebiasis.

Acknowledgments

Research in the authors' laboratories was supported in part by Grant 254434 (to CR) from Consejo Nacional de Ciencia y Tecnología (CONACyT), Mexico (http://www.conacyt.gob.mx)

Conflict of Interest

The authors declare not having any commercial or financial relationships that could be interpreted as a potential conflict of interest

References

Abd-Alla, M.D., Jackson, T.F., Rogers, T., Reddy, S., and Ravdin, J.I. (2006). Mucosal immunity to asymptomatic *Entamoeba histolytica* and *Entamoeba dispar* infection is associated with a peak intestinal anti-lectin immunoglobulin A antibody response. *Infect. Immun.* 74: 3897-3903. doi: 10.1128/IAI.02018-05.

Abhyankar, M.M., Orr, M.T., Lin, S., Suraju, M.O., Simpson, A., Blust, M., et al. (2018). Adjuvant composition and delivery route shape immune response quality and protective efficacy of a recombinant vaccine for *Entamoeba histolytica. NPJ Vaccines* 3: 22. doi: 10.1038/s41541-018-0060-x.

Acuna-Soto, R., Maguire, J.H., and Wirth, D.F. (2000). Gender distribution in asymptomatic and invasive amebiasis. *Am. J, Gastroenterol.* 95: 1277-1283. doi: 10.1111/j.1572-0241.2000.01525.x.

Agarwal, S., Anand, G., Sharma, S., Parimita Rath, P., Gourinath, S., and Bhattacharya, A. (2019). EhP3, a homolog of 14-3-3 family of protein participates in actin reorganization and phagocytosis in *Entamoeba histolytica. PLoS Pathog.* 15: e1007789. doi: 10.1371/journal.ppat. 1007789.

Aguirre García, M., Gutiérrez-Kobeh, L., and López Vancell, R. (2015). *Entamoeba histolytica*: adhesins and lectins in the trophozoite surface. *Molecules* 20: 2802-2815. doi: 10.3390/molecules20022802.

Andrä, J., Herbst, R., and Leippe, M. (2003). Amoebapores, archaic effector peptides of protozoan origin, are discharged into phagosomes and kill bacteria by permeabilizing their membranes. *Dev. Comp. Immunol.* 27: 291-304. doi: 10.1016/s0145-305x(02)00106-4.

Ankri, S., Padilla-Vaca, F., Stolarsky, T., Koole, L., Katz, U., and Mirelman, D. (1999). Antisense inhibition of expression of the light subunit (35 kDa) of the Gal/GalNac lectin complex inhibits *Entamoeba histolytica* virulence. *Mol. Microbiol.* 33: 327-337. doi: 10.1046/j. 1365-2958.1999.01476.x.

Avalos-Padilla, Y., Knorr, R.L., Javier-Reyna, R., García-Rivera, G., Lipowsky, R., Dimova, R., et al. (2018). The conserved ESCRT-III machinery participates in the phagocytosis of *Entamoeba histolytica*. *Front. Cell. Infect. Microbiol.* 8: 53. doi: 10.3389/fcimb.2018.00053.

Ávila, E.E., Salaiza, N., Pulido, J., Rodríguez, M.C., Díaz-Godínez, C., Laclette, J.P., et al. (2016). *Entamoeba histolytica* trophozoites and lipopeptidophosphoglycan trigger human neutrophil extracellular traps. *PLos One* 11: e0158979. doi: 10.1371/journal.pone.0158979.

Balderas-Renteria, I., García-Lázaro, J.F., Carranza-Rosales, P., Morales-Ramos, L.H., Galan-Wong, L.J., and Muñoz-Espinosa, L.E. (2007). Transcriptional upregulation of genes related to virulence activation in *Entamoeba histolytica*. *Arch. Med. Res.* 38: 372-379. doi: 10.1016/j.arcmed.2007.01.003.

Bansal, D., Ave, P., Kerneis, S., Frileux, P., Boché, O., Baglin, A.C., et al. (2009). An ex-vivo human intestinal model to study *Entamoeba histolytica* pathogenesis. *PLoS Negl. Trop. Dis.* 3: e551. doi: 10.1371/journal.pntd.0000551.

Bansal, D., Sehgal, R., Chawla, Y., Malla, N., and Mahajan, R.C. (2005). Cytokine mRNA expressions in symptomatic vs. asymptomatic amoebiasis patients. *Parasite Immunol.* 27: 37-43. doi: 10.1111/j.1365-3024.2005.00739.x.

Baxt, L.A., Baker, R.P., Singh, U., and Urban, S. (2008). An *Entamoeba histolytica* rhomboid protease with atypical specificity cleaves a surface lectin involved in phagocytosis and immune evasion. *Genes Dev.* 22: 1636-1346. doi: 10.1101/gad.1667708.

Baxt, L.A., Rastew, E., Bracha, R., Mirelman, D., and Singh, U. (2010). Downregulation of an *Entamoeba histolytica* rhomboid protease reveals roles in regulating parasite adhesion and phagocytosis. *Eukaryot. Cell* 9: 1283-1293. doi: 10.1128/EC.00015-10.

Bernin, H., Marggraff, C., Jacobs, T., Brattig, N., Le, V.A., Blessmann, J., et al. (2014). Immune markers characteristic for asymptomatically infected and diseased *Entamoeba histolytica* individuals and their relation to sex. *BMC Infect. Dis.* 14: 621. doi: 10.1186/s12879-014-0621-1.

Bharadwaj, R., Sharma, S., Janhawi Arya, R., Bhattacharya, S., and Bhattacharya, A. (2018). EhRho1 regulates phagocytosis by modulating actin dynamics through EhFormin1 and EhProfilin1 in *Entamoeba histolytica. Cell. Microbiol.* 20: e12851. doi: 10.1111/cmi. 12851.

Bhattacharya, A., Arya, R., Clark, C.G., and Ackers, J.P. (2000). Absence of lipophosphoglycan-like glycoconjugates in *Entamoeba dispar. Parasitology* 120: 31-35. doi: 10.1017/s0031182099005259.

Biller, L., Matthiesen, J., Kühne, V., Lotter, H., Handal, G., Nozaki, T., et al. (2014). The cell surface proteome of *Entamoeba histolytica. Mol. Cell. Proteomics* 13: 132-144. doi: 10.1074/mcp.M113.031393.

Birchenough, G.M., Johansson, M.E., Gustafsson, J.K., Bergström, J.H., and Hansson, G.C. (2015). New developments in goblet cell mucus secretion and function. *Mucosal Immunol.* 8: 712-719. doi: 10.1038/mi.2015.32.

Blazquez, S., Rigothier, M.C., Huerre, M., and Guillén, N. (2007). Initiation of inflammation and cell death during liver abscess formation by *Entamoeba histolytica* depends on activity of the galactose/N-acetyl-D-galactosamine lectin. *Int. J. Parasitol.* 37: 425-433. doi: 10.1016/j.ijpara.2006.10.008.

Braga, L.L., Ninomiya, H., McCoy, J.J., Eacker, S., Wiedmer, T., Pham, C., et al. (1992). Inhibition of the complement membrane attack complex by the galactose-specific adhesion of *Entamoeba histolytica. J. Clin. Invest.* 90: 1131-1137. doi: 10.1172/JCI115931.

Branzk, N., Lubojemska, A., Hardison, S.E., Wang, Q., Gutierrez, M.G., Brown, G.D., et al. (2014). Neutrophils sense microbe size and selectively release neutrophil extracellular traps in response to large pathogens. *Nat. Immunol.* 15: 1017–1025. doi:10.1038/ni.2987.

Bruhn, H., and Leippe, M. (2001). Novel putative saposin-like proteins of *Entamoeba histolytica* different from amoebapores. *Biochim. Biophys. Acta* 1514: 14-20. doi: 10.1016/S0005-2736(01)00345-5.

Burgess, S.L., and Petri, W.A., Jr (2016). The intestinal bacterial microbiome and *E. histolytica* infection. *Curr. Trop. Med. Rep.* 3: 71-74. doi: 10.1007/s40475-016-0083-1.

Campos-Rodríguez, R., Gutiérrez-Meza, M., Jarillo-Luna, R.A., Drago-Serrano, M.E., Abarca-Rojano, E., Ventura-Juárez, J., et al. (2016). A review of the proposed role of neutrophils in rodent amebic liver abscess models. *Parasite* 23: 6. doi: 10.1051/parasite/2016006.

Campos-Rodríguez, R., and Jarillo-Luna, A. (2005). The pathogenicity of *Entamoeba histolytica* is related to the capacity of evading innate immunity. *Parasite Immunol.* 27: 1-8. doi: 10.1111/j.1365-3024.2005.00743.x.

Cao, A.T., Yao, S., Gong, B., Elson, C.O., and Cong, Y. (2012). Th17 cells upregulate polymeric Ig receptor and intestinal IgA and contribute to intestinal homeostasis. *J. Immunol.* 189: 4666-4673. doi: 10.4049/jimmunol.1200955.

Choi, M.H., Sajed, D., Poole, L., Hirata, K., Herdman, S., Torian, B.E., et al. (2005). An unusual surface peroxiredoxin protects invasive *Entamoeba histolytica* from oxidant attack. *Mol. Biochem. Parasitol.* 143: 80-89. doi: 10.1016/j.molbiopara.2005.04.014.

Christy, N.C., and Petri, W.A., Jr (2011). Mechanisms of adherence, cytotoxicity and phagocytosis modulate the pathogenesis of *Entamoeba histolytica*. *Future Microbiol.* 6: 1501-1519. doi: 10.2217/fmb.11.120.

Constantino-Jonapa, L.A., Hernández-Ramírez, V.I., Osorio-Trujillo, C., and Talamás-Rohana, P. (2018). EhRab21 mobilization during erythrophagocytosis in *Entamoeba histolytica*. *Microsc. Res. Tech.* 81: 1024-1035. doi: 10.1002/jemt.23069.

Cornick, S., and Chadee, K. (2017). *Entamoeba histolytica*: Host parasite interactions at the colonic epithelium. *Tissue Barriers* 5: e1283386. doi: 10.1080/21688370.2017.1283386.

Cornick, S., Moreau, F., and Chadee, K. (2016). *Entamoeba histolytica* cysteine proteinase 5 evokes mucin exocytosis from colonic goblet cells via αvβ3 integrin. *PLoS Pathog.* 12: e1005579. doi: 10.1371/journal.ppat.1005579.

Cuellar, P., Hernández-Nava, E., García-Rivera, G., Chávez-Munguía, B., Schnoor, M., Betanzos, A., et al. (2017). *Entamoeba histolytica* EhCP112 dislocates and degrades claudin-1 and claudin-2 at tight junctions of the intestinal epithelium. *Front. Cell. Infect. Microbiol.* 7: 372. doi: 10.3389/fcimb.2017.00372.

Davis, P.H., Schulze, J., and Stanley, S.L.J. (2007). Transcriptomic comparison of two *Entamoeba histolytica* strains with defined virulence phenotypes identifies new virulence factor candidates and key differences in the expression patterns of cysteine proteases, lectin light chains, and calmodulin. *Mol. Biochem. Parasitol.* 151: 118-128. doi: 10.1016/j.molbiopara.2006.10.014.

Davis, P.H., Zhang, X., Guo, J., Townsend, R.R., and Stanley, S.L., Jr (2006). Comparative proteomic analysis of two *Entamoeba histolytica* strains with different virulence phenotypes identifies peroxiredoxin as an important component of amoebic virulence. *Mol. Microbiol.* 61: 1523-1532. doi: 10.1111/j.1365-2958.2006.05344.x.

Deloer, S., Nakamura, R., Kikuchi, M., Moriyasu, T., Kalenda, Y.D.J., Mohammed, E.S., et al. (2017). IL-17A contributes to reducing IFN-γ/IL-4 ratio and persistence of *Entamoeba histolytica* during intestinal amebiasis. *Parasitol. Int.* 66: 817-823. doi: 10.1016/j.parint.2017.09.011.

Dey, I., and Chadee, K. (2008). Prostaglandin E2 produced by *Entamoeba histolytica* binds to EP4 receptors and stimulates interleukin-8 production in human colonic cells. *Infect. Immun.* 76: 5158-5163. doi: 10.1128/IAI.00645-08.

Díaz-Godínez, C., Fonseca, Z., Néquiz, M., Laclette, J.P., Rosales, C., and Carrero, J.C. (2018). *Entamoeba histolytica* trophozoites induce a rapid non-classical NETosis mechanism independent of NOX2-derived reactive oxygen species and PAD4 activity. *Front. Cell. Infect. Microbiol.* 8: 184. doi: 10.3389/fcimb.2018.00184.

Dickson-Gonzalez, S.M., de Uribe, M.L., and Rodriguez-Morales, A.J. (2009). Polymorphonuclear neutrophil infiltration intensity as consequence of *Entamoeba histolytica* density in amebic colitis. *Surg. Infect.* 10: 91-97. doi: 10.1089/sur.2008.011.

Dolabella, S.S., Serrano-Luna, J., Navarro-García, F., Cerritos, R., Ximénez, C., Galván-Moroyoqui, J.M., et al. (2012). Amoebic liver abscess production by *Entamoeba dispar*. *Ann. Hepatol.* 11: 107-117.

Duggal, P., Guo, X., Haque, R., Peterson, K.M., Ricklefs, S., Mondal, D., et al. (2011). A mutation in the leptin receptor is associated with

Entamoeba histolytica infection in children. *J. Clin. Invest.* 121: 1191-1198. doi: 10.1172/JCI45294.

Duggal, P., Haque, R., Roy, S., Mondal, D., Sack, R.B., Farr, B.M., et al. (2004). nfluence of human leukocyte antigen class II alleles on susceptibility to *Entamoeba histolytica* infection in Bangladeshi children. *J. Infect. Dis.* 189: 520-526. doi: 10.1086/381272.

Duplessis, C.A., Gutierrez, R.L., and Porter, C.K. (2017). Review: chronic and persistent diarrhea with a focus in the returning traveler. *Trop. Dis. Travel Med. Vaccines.* 3: 9. doi: 10.1186/s40794-017-0052-2.

Elnekave, K., Siman-Tov, R., and Ankri, S. (2003). Consumption of L-arginine mediated by *Entamoeba histolytica* L-arginase (EhArg) inhibits amoebicidal activity and nitric oxide production by activated macrophages. *Parasite Immunol.* 25: 597-608. doi: 10.1111/j.0141-9838.2004.00669.x.

Espinosa-Cantellano, M., and Martínez-Palomo, A. (2000). Pathogenesis of intestinal amebiasis: from molecules to disease. *Clin. Microbiol. Rev.* 13: 318-331. doi: 10.1128/CMR.13.2.318-331.2000.

Estrada-Figueroa, L.A., Ramírez-Jiménez, Y., Osorio-Trujillo, C., Shibayama, M., Navarro-García, F., García-Tovar, C., et al. (2011). Absence of CD38 delays arrival of neutrophils to the liver and innate immune response development during hepatic amoebiasis by *Entamoeba histolytica*. *Parasite Immunol.* 33: 661-668. doi: 10.1111/j.1365-3024.2011.01333.x.

Fonseca, Z., Díaz-Godínez, C., Mora, N., Alemán, O.R., Uribe-Querol, E., Carrero, J.C., et al. (2018). *Entamoeba histolytica* induce signaling via Raf/MEK/ERK for neutrophil extracellular trap (NET) formation. *Front. Cell. Infect. Microbiol.* 8: 226. doi: 10.3389/fcimb.2018.00226.

Fonseca, Z., Uribe-Querol, E., Díaz-Godínez, C., Carrero, J.C., and Rosales, C. (2019). Pathogenic *Entamoeba histolytica*, but not *Entamoeba dispar,* induce neutrophil extracellular trap (NET) formation. *J. Leukoc. Biol.* 105: 1167-1181. doi: 10.1002/JLB. MA0818-309RRR.

Fotedar, R., Stark, D., Beebe, N., Marriott, D., Ellis, J., and Harkness, J. (2007). Laboratory diagnostic techniques for *Entamoeba* species. *Clin. Microbiol. Rev.* 20: 511-532. doi: 10.1128/CMR.00004-07.

Frederick, J.R., and Petri, W.A., Jr (2005). Roles for the galactose-/N-acetylgalactosamine-binding lectin of *Entamoeba* in parasite virulence and differentiation. *Glycobiology* 15: 53R-59R. doi: 10.1093/glycob/cwj007.

Freitas, M.A., Fernandes, H.C., Calixto, V.C., Martins, A.S., Silva, E.F., Pesquero, J.L., et al. (2009). *Entamoeba histolytica*: cysteine proteinase activity and virulence. Focus on cysteine proteinase 5 expression levels. *Exp. Parasitol.* 122: 306-309. doi: 10.1016/j.exppara.2009.04.005.

Galván-Moroyoqui, J.M., Domínguez-Robles, M.d.C., and Meza, I. (2011). Pathogenic bacteria prime the induction of Toll-like receptor signalling in human colonic cells by the Gal/GalNAc lectin carbohydrate recognition domain of *Entamoeba histolytica*. *Int. J. Parasitol.* 41: 1101-1112. doi: 10.1016/j.ijpara.2011.06.003.

Garcia-Nieto, R.M., Rico-Mata, R., Arias-Negrete, S., and Avila, E.E. (2008). Degradation of human secretory IgA1 and IgA2 by *Entamoeba histolytica* surface-associated proteolytic activity. *Parasitol. Int.* 57: 417-423. doi: 10.1016/j.parint.2008.04.013.

García-Zepeda, E.A., Rojas-López, A., Esquivel-Velázquez, M., and Ostoa-Saloma, P. (2007). Regulation of the inflammatory immune response by the cytokine/chemokine network in amoebiasis. *Parasite Immunol.* 29: 679-684. doi: 10.1111/j.1365-3024.2007.00990.x.

Ghadirian, E., and Denis, M. (1992). In vivo activation of macrophages by IFN-gamma to kill *Entamoeba histolytica* trophozoites in vitro. *Parasite Immunol.* 14: 397-404. doi: 10.1111/j.1365-3024.1992.tb00014.x.

Ghosh, A.S., Dutta, S., and Raha, S. (2010). Hydrogen peroxide-induced apoptosis-like cell death in *Entamoeba histolytica*. *Parasitol. Int.* 59: 166-172. doi: 10.1016/j.parint.2010.01.001.

Ghosh, S., Padalia, J., and Moonah, S. (2019). Tissue destruction caused by *Entamoeba histolytica* parasite: Cell death, inflammation, invasion, and the gut microbiome. *Curr. Clin. Microbiol. Rep.* 6: 51-57. doi: 10.1007/s40588-019-0113-6.

Gilchrist, C.A., Petri, S.E., Schneider, B.N., Reichman, D.J., Jiang, N., Begum, S., et al. (2016). Role of the gut microbiota of children in diarrhea due to the protozoan parasite *Entamoeba histolytica*. *J. Infect. Dis.* 213: 1579-1585. doi: 10.1093/infdis/jiv772.

Gilmartin, A.A., Ralston, K.S., and Petri, W.A., Jr (2017). Inhibition of amebic lysosomal acidification blocks amebic trogocytosis and cell killing. *mBio* 8: e01187-01117. doi: 10.1128/mBio.01187-17.

Gonzalez Rivas, E., Ximenez, C., Nieves-Ramirez, M.E., Moran Silva, P., Partida-Rodríguez, O., Hernandez, E.H., et al. (2018). *Entamoeba histolytica* calreticulin induces the expression of cytokines in peripheral blood mononuclear cells isolated from patients with amebic liver abscess. *Front. Cell. Infect. Microbiol.* 8: 358. doi: 10.3389/fcimb.2018.00358.

Guha-Niyogi, A., Sullivan, D.R., and Turco, S.J. (2001). Glycoconjugate structures of parasitic protozoa. *Glycobiology* 11: 45R-59R. doi: 10.1093/glycob/11.4.45r.

Guillén, N. (2019). The interaction between *Entamoeba histolytica* and enterobacteria shed light on an ancient antibacterial response. *Cell. Microbiol.* 21: e13039. doi:10.1111/cmi.13039.

Guo, X., Barroso, L., Becker, S.M., Lyerly, D.M., Vedvick, T.S., Reed, S.G., et al. (2009). Protection against intestinal amebiasis by a recombinant vaccine is transferable by T cells and mediated by gamma interferon. *Infect. Immun.* 77: 3909-3918. doi: 10.1128/IAI.00487-09.

Guo, X., Barroso, L., Lyerly, D.M., Petri, W.A., Jr, and Houpt, E.R. (2011a). $CD4^+$ and $CD8^+$ T cell- and IL-17-mediated protection against *Entamoeba histolytica* induced by a recombinant vaccine. *Vaccine* 29: 772-777. doi: 10.1016/j.vaccine.2010.11.013.

Guo, X., Roberts, M.R., Becker, S.M., Podd, B., Zhang, Y., Chua, S.C., Jr, et al. (2011b). Leptin signaling in intestinal epithelium mediates resistance to enteric infection by *Entamoeba histolytica*. *Mucosal Immunol.* 4: 294-303. doi: 10.1038/mi.2010.76.

Guo, X., Stroup, S.E., and Houpt, E.R. (2008). Persistence of *Entamoeba histolytica* infection in CBA mice owes to intestinal IL-4 production and inhibition of protective IFN-gamma. *Mucosal Immunol.* 1: 139-146. doi: 10.1038/mi.2007.18.

Gutierrez, M.G. (2013). Functional role(s) of phagosomal Rab GTPases. *Small GTPases* 4: 148-158. doi: 10.4161/sgtp.25604.

Haghighi, A., Kobayashi, S., Takeuchi, T., Masuda, G., and Nozaki, T. (2002). Remarkable genetic polymorphism among *Entamoeba histolytica* isolates from a limited geographic area. *J. Clin. Microbiol.* 40: 4081-4090. doi: 10.1128/jcm.40.11.4081-4090.2002.

Hamano, S., Asgharpour, A., Stroup, S.E., Wynn, T.A., Leiter, E.H., and Houpt, E. (2006). Resistance of C57BL/6 mice to amoebiasis is mediated by nonhemopoietic cells but requires hemopoietic IL-10 production. *J. Immunol.* 177: 1208-1213. doi: 10.4049/jimmunol.177.2.1208.

Haque, R., Ali, I.M., Sack, R.B., Farr, B.M., Ramakrishnan, G., and Petri, W.A., Jr (2001). Amebiasis and mucosal IgA antibody against the *Entamoeba histolytica* adherence lectin in Bangladeshi children. *J. Infect. Dis.* 183: 1787-1793. doi: 10.1086/320740.

Haque, R., Duggal, P., Ali, I.M., Hossain, M.B., Mondal, D., Sack, R.B., et al. (2002). Innate and acquired resistance to amebiasis in Bangladeshi children. *J. Infect. Dis.* 186: 547-552. doi: 10.1086/341566.

Haque, R., Huston, C.D., Hughes, M., Houpt, E., and Petri, W.A., Jr (2003). Amebiasis. *N. Engl. J. Med.* 348: 1565-1573. doi: 10.1056/NEJMra022710.

Haque, R., Kress, K., Wood, S., Jackson, T.F., Lyerly, D., Wilkins, T., et al. (1993). Diagnosis of pathogenic *Entamoeba histolytica* infection using a stool ELISA based on monoclonal antibodies to the galactose-specific adhesin. *J. Infect. Dis.* 167: 247-249. doi: 10.1093/infdis/167.1.247.

Haque, R., Mondal, D., Duggal, P., Kabir, M., Roy, S., Farr, B.M., et al. (2006). *Entamoeba histolytica* infection in children and protection from subsequent amebiasis. *Infect. Immun.* 74: 904-909. doi: 10.1128/IAI.74.2.904-909.2006.

Haque, R., Mondal, D., Shu, J., Roy, S., Kabir, M., Davis, A.N., et al. (2007). Correlation of interferon-gamma production by peripheral blood mononuclear cells with childhood malnutrition and susceptibility to amebiasis. *Am. J. Trop. Med. Hyg.* 76: 340-344.

Helk, E., Bernin, H., Ernst, T., Ittrich, H., Jacobs, T., Heeren, J., et al. (2013). TNFα-mediated liver destruction by Kupffer cells and Ly6Chi monocytes during *Entamoeba histolytica* infection. *PLoS Pathog.* 9: e1003096. doi: 10.1371/journal.ppat.1003096.

Hernández, E.G., Granados, J., Partida-Rodríguez, O., Valenzuela, O., Rascón, E., Magaña, U., et al. (2015). Prevalent HLA class II alleles in Mexico City appear to confer resistance to the development of amebic liver abscess. *PLoS One* 10: e0126195. doi: 10.1371/journal.pone.0126195.

Hernández-Cuevas, N.A., Jhingan, G.D., Petropolis, D., Vargas, M., and Guillen, N. (2019). Acetylation is the most abundant actin modification in *Entamoeba histolytica* and modifications of actin's amino-terminal domain change cytoskeleton activities. *Cell. Microbiol.* 21: e12983. doi: 10.1111/cmi.12983.

Herricks, J.R., Hotez, P.J., Wanga, V., Coffeng, L.E., Haagsma, J.A., Basáñez, M.G., et al. (2017). The global burden of disease study 2013: What does it mean for the NTDs? *PLoS Negl. Trop. Dis.* 11: e0005424. doi: 10.1371/journal.pntd.0005424.

Houpt, E.R., Glembocki, D.J., Obrig, T.G., Moskaluk, C.A., Lockhart, L.A., Wright, R.L., et al. (2002). The mouse model of amebic colitis reveals mouse strain susceptibility to infection and exacerbation of disease by $CD4^+$ T cells. *J. Immunol.* 169: 4496-4503. doi: 10.4049/jimmunol.169.8.4496.

Hung, C.C., Ji, D.D., Sun, H.Y., Lee, Y.T., Hsu, S.Y., Chang, S.Y., et al. (2008). Increased risk for *Entamoeba histolytica* infection and invasive amebiasis in HIV seropositive men who have sex with men in Taiwan. *PLoS Negl. Trop. Dis.* 2: e175. doi: 10.1371/journal.pntd.0000175.

Huston, C.D., Boettner, D.R., Miller-Sims, V., and Petri, W.A., Jr (2003). Apoptotic killing and phagocytosis of host cells by the parasite

Entamoeba histolytica. *Infect. Immun.* 71: 964-972. doi: 10.1128/iai. 71.2.964-972.2003.

Huston, C.D., Houpt, E.R., Mann, B.J., Hahn, C.S., and Petri, W.A., Jr (2000). Caspase 3-dependent killing of host cells by the parasite *Entamoeba histolytica*. *Cell. Microbiol.* 2: 617-625. doi: 10.1046/j. 1462-5822.2000.00085.x.

Irmer, H., Tillack, M., Biller, L., Handal, G., Leippe, M., Roeder, T., et al. (2009). Major cysteine peptidases of *Entamoeba histolytica* are required for aggregation and digestion of erythrocytes but are dispensable for phagocytosis and cytopathogenicity. *Mol. Microbiol.* 72: 658-667. doi: 10.1111/j.1365-2958.2009.06672.x.

Iyer, L.R., Verma, A.K., Paul, J., and Bhattacharya, A. (2019). Phagocytosis of gut bacteria by *Entamoeba histolytica*. *Front. Cell. Infect. Microbiol.* 9: 34. doi: 10.3389/fcimb.2019.00034.

Jarillo-Luna, R.A., Campos-Rodríguez, R., and Tsutsumi, V. (2002). *Entamoeba histolytica*: Immunohistochemical study of hepatic amoebiasis in mouse. Neutrophils and nitric oxide as possible factors of resistance. *Exp. Parasitol.* 101: 40-56. doi: 10.1016/S0014-4894 (02)00021-8.

Javier-Reyna, R., Montaño, S., García-Rivera, G., Rodríguez, M.A., González-Robles, A., and Orozco, E. (2019). EhRabB mobilises the EhCPADH complex through the actin cytoskeleton during phagocytosis of *Entamoeba histolytica*. *Cell. Microbiol.* 21: e13071. doi: 10.1111/cmi.13071.

Kammanadiminti, S.J., Mann, B.J., Dutil, L., and Chadee, K. (2004). Regulation of Toll-like receptor-2 expression by the Gal-lectin of *Entamoeba histolytica*. *FASEB J.* 18: 155-157. doi: 10.1096/fj.03-0578fje.

Kantor, M., Abrantes, A., Estevez, A., Schiller, A., Torrent, J., Gascon, J., et al. (2018). *Entamoeba histolytica*: Updates in clinical manifestation, pathogenesis, and vaccine development. *Can. J. Gastroenterol. Hepatol.* 2018: 4601420. doi: 10.1155/2018/4601420.

Katz, U., Ankri, S., Stolarsky, T., Nuchamowitz, Y., and Mirelman, D. (2002). *Entamoeba histolytica* expressing a dominant negative Ntruncated light subunit of its gal-lectin are less virulent. *Mol. Biol. Cell* 13: 4256-4265. doi: 10.1091/mbc.e02-06-0344.

Krishnan, D., and Ghosh, S.K. (2018). Cellular events of multinucleated giant cells formation during the encystation of *Entamoeba invadens*. *Front. Cell. Infect. Microbiol.* 8: 262. doi: 10.3389/fcimb.2018.00262.

Labruyère, E., and Guillén, N. (2006). Host tissue invasion by *Entamoeba histolytica* is powered by motility and phagocytosis. *Arch. Med. Res.* 37: 253-258. doi: 10.1016/j.arcmed.2005.10.005.

Labruyère, E., Thibeaux, R., Olivo-Marin, J.C., and Guillén, N. (2019). Crosstalk between *Entamoeba histolytica* and the human intestinal tract during amoebiasis. *Parasitology* 146: 1140-1149. doi: 10.1017/S0031182017002190.

Lejeune, M., Moreau, F., and Chadee, K. (2011). Prostaglandin E2 produced by *Entamoeba histolytica* signals via EP4 receptor and alters claudin-4 to increase ion permeability of tight junctions. *Am. J. Pathol.* 179: 807-818. doi: 10.1016/j.ajpath.2011.05.001.

Leon-Coria, A., Kumar, M., and Chadee, K. (2020). The delicate balance between *Entamoeba histolytica*, mucus and microbiota. *Gut Microbes* 11: 118-125. doi: 10.1080/19490976.2019.1614363.

Leon-Coria, A., Kumar, M., Moreau, F., and Chadee, K. (2018). Defining cooperative roles for colonic microbiota and Muc2 mucin in mediating innate host defense against *Entamoeba histolytica*. *PLoS Pathog.* 14: e1007466. doi: 10.1371/journal.ppat.1007466.

Levin, R., Grinstein, S., and Canton, J. (2016). The life cycle of phagosomes: formation, maturation, and resolution. *Immunol. Rev.* 273: 156-179. doi: 10.1111/imr.12439.

Liang, S.C., Tan, X.Y., Luxenberg, D.P., Karim, R., Dunussi-Joannopoulos, K., Collins, M., et al. (2006). Interleukin (IL)-22 and IL-17 are coexpressed by Th17 cells and cooperatively enhance expression of antimicrobial peptides. *J. Exp. Med.* 203: 2271-2279. doi: 10.1084/jem.20061308.

Lidell, M.E., Moncada, D.M., Chadee, K., and Hansson, G.C. (2006). *Entamoeba histolytica* cysteine proteases cleave the MUC2 mucin in its C-terminal domain and dissolve the protective colonic mucus gel.

Proc. Natl. Acad. Sci. U.S.A. 103: 9298-9303. doi: 10.1073/pnas. 0600623103.

Lin, J.Y., and Chadee, K. (1992). Macrophage cytotoxicity against *Entamoeba histolytica* trophozoites is mediated by nitric oxide from L-arginine. *J. Immunol.* 148: 3999-4005.

López-López, P., Martínez-López, M.C., Boldo-León, X.M., Hernández-Díaz, Y., González-Castro, T.B., Tovilla-Zárate, C.A., et al. (2017). Detection and differentiation of *Entamoeba histolytica* and *Entamoeba dispar* in clinical samples through PCR-denaturing gradient gel electrophoresis. *Braz. J. Med. Biol. Res.* 50: e5997. doi: 10.1590/1414-431X20175997.

Lotter, H., González-Roldán, N., Lindner, B., Winau, F., Isibasi, A., Moreno-Lafont, M., et al. (2009). Natural killer T cells activated by a lipopeptidophosphoglycan from *Entamoeba histolytica* are critically important to control amebic liver abscess. *PLoS Pathog.* 5: e1000434. doi: 10.1371/journal.ppat.1000434.

Lotter, H., Helk, E., Bernin, H., Jacobs, T., Prehn, C., Adamski, J., et al. (2013). Testosterone increases susceptibility to amebic liver abscess in mice and mediates inhibition of IFNγ secretion in natural killer T cells. *PLoS One* 8: e55694. doi: 10.1371/journal.pone.0055694.

Mackey-Lawrence, N.M., Guo, X., Sturdevant, D.E., Virtaneva, K., Hernandez, M.M., Houpt, E., et al. (2013). Effect of the leptin receptor Q223R polymorphism on the host transcriptome following infection with *Entamoeba histolytica*. *Infect. Immun.* 81: 1460-1470. doi: 10.1128/IAI.01383-12.

Maldonado-Barrera, C.A., Campos-Esparza, M.R., Muñoz-Fernández, L., Victoria-Hernández, J.A., Campos-Rodríguez, R., Talamás-Rohana, P., et al. (2012). Clinical case of cerebral amebiasis caused by *E. histolytica*. *Parasitol. Res.* 110: 1291-1296. doi: 10.1007/s00436-011-2617-8.

Maldonado-Bernal, C., Kirschning, C.J., Rosenstein, Y., Rocha, L.M., Rios-Sarabia, N., Espinosa-Cantellano, M., et al. (2005). The innate immune response to *Entamoeba histolytica* lipopeptidophosphoglycan is mediated by toll-like receptors 2 and 4. *Parasite Immunol.* 27: 127-137. doi: 10.1111/j.1365-3024.2005.00754.x.

Manich, M., Hernandez-Cuevas, N., Ospina-Villa, J.D., Syan, S., Marchat, L.A., Olivo-Marin, J.C., et al. (2018). Morphodynamics of the actin-rich cytoskeleton in *Entamoeba histolytica*. *Front. Cell. Infect. Microbiol.* 8: 179. doi: 10.3389/fcimb.2018.00179.

Mao, Y., and Finnemann, S.C. (2015). Regulation of phagocytosis by Rho GTPases. *Small GTPases* 6: 89-99. doi: 10.4161/21541248.2014.989785.

Marie, C., and Petri, W.A., Jr (2014). Regulation of virulence of *Entamoeba histolytica*. *Annu. Rev. Microbiol.* 68: 493-520. doi: 10.1146/annurev-micro-091313-103550.

Marie, C., Verkerke, H.P., Theodorescu, D., and Petri, W.A., Jr (2015). A whole-genome RNAi screen uncovers a novel role for human potassium channels in cell killing by the parasite *Entamoeba histolytica*. *Sci. Rep.* 5: 13613. doi: 10.1038/srep13613.

Mayadas, T.N., Cullere, X., and Lowell, C.A. (2014). The multifaceted functions of neutrophils. *Annu. Rev. Pathol.* 9: 181-218 doi: 10.1146/annurev-pathol-020712-164023.

Miller, H.W., Suleiman, R.L., and Ralston, K.S. (2019). Trogocytosis by *Entamoeba histolytica* mediates acquisition and display of human cell membrane proteins and evasion of lysis by human serum. *mBio* 10: e00068-00019. doi: 10.1128/mBio.00068-19.

Moncada, D., Keller, K., and Chadee, K. (2005). *Entamoeba histolytica*-secreted products degrade colonic mucin oligosaccharides. *Infect. Immun.* 73: 3790-3793. doi: 10.1128/IAI.73.6.3790-3793.2005.

Moonah, S.N., Jiang, N.M., and Petri, W.A., Jr (2013). Host immune response to intestinal amebiasis. *PLoS Pathog.* 9: e1003489. doi: 10.1371/journal.ppat.1003489.

Mortimer, L., and Chadee, K. (2010). The immunopathogenesis of *Entamoeba histolytica*. *Exp. Parasitol.* 126: 366-380. doi: 10.1016/j.exppara.2010.03.005.

Nakada-Tsukui, K., and Nozaki, T. (2016). Immune response of amebiasis and immune evasion by *Entamoeba histolytica*. *Front. Immunol.* 7: 175. doi: 10.3389/fimmu.2016.00175.

Neeli, I., and Radic, M. (2012). Knotting the NETs: analyzing histone modifications in neutrophil extracellular traps. *Arthritis Res. Ther.* 14: 115. doi: 10.1186/ar3773.

Ngobeni, R., Abhyankar, M.M., Jiang, N.M., Farr, L.A., Samie, A., Haque, R., et al. (2017a). *Entamoeba histolytica*-encoded homolog of macrophage migration inhibitory factor contributes to mucosal inflammation during amebic colitis. *J. Infect. Dis.* 215: 1294-1302. doi: 10.1093/infdis/jix076.

Ngobeni, R., Samie, A., Moonah, S., Watanabe, K., Petri, W.A., Jr, and Gilchrist, C. (2017b). *Entamoeba* species in South Africa: Correlations with the host microbiome, parasite burdens, and first description of *Entamoeba bangladeshi* outside of Asia. *J. Infect. Dis.* 216: 1592-1600. doi: 10.1093/infdis/jix535.

Nishida, A., Lau, C.W., Zhang, M., Andoh, A., Shi, H.N., Mizoguchi, E., et al. (2012). The membrane-bound mucin Muc1 regulates T helper 17-cell responses and colitis in mice. *Gastroenterology* 142: 865-874. doi: 10.1053/j.gastro.2011.12.036.

Nunes, M.C., Guimarães Júnior, M.H., Diamantino, A.C., Gelape, C.L., and Ferrari, T.C. (2017). Cardiac manifestations of parasitic diseases. *Heart* 103: 651-658. doi: 10.1136/heartjnl-2016-309870.

Olivos-García, A., Carrero, J.C., Ramos, E., Nequiz, M., Tello, E., Montfort, I., et al. (2007). Late experimental amebic liver abscess in hamster is inhibited by cyclosporine and N-acetylcysteine. *Exp. Mol. Pathol.* 82: 310-315. doi: 10.1016/j.yexmp.2006.09.005.

Orozco, E., Guarneros, G., Martinez-Palomo, A., and T, S. (1983). *Entamoeba histolytica*. Phagocytosis as a virulence factor. *J. Exp. Med.* 158. 1511-1521. doi: 10.1084/jem.158.5.1511.

Papayannopoulos, V., and Zychlinsky, A. (2009). NETs: a new strategy for using old weapons. *Trends. Immunol.* 30: 513-521. doi: 10.1016/j.it.2009.07.011.

Pauwels, A.M., Trost, M., Beyaert, R., and Hoffmann, E. (2017). Patterns, receptors, and signals: Regulation of phagosome maturation. *Trends Immunol.* 38: 407-422. doi: 10.1016/j.it.2017.03.006.

Pelaseyed, T., Bergström, J.H., Gustafsson, J.K., Ermund, A., Birchenough, G.M., Schütte, A., et al. (2014). The mucus and mucins of the goblet cells and enterocytes provide the first defense line of the gastrointestinal tract and interact with the immune system. *Immunol. Rev.* 260: 8-20. doi: 10.1111/imr.12182.

Pérez-Tamayo, R., Montfort, I., García, A.O., Ramos, E., and Ostria, C.B. (2006). Pathogenesis of acute experimental liver amebiasis. *Arch. Med. Res.* 37: 203-209. doi: 10.1016/j.arcmed.2005.10.007.

Peterson, K.M., Shu, J., Duggal, P., Haque, R., Mondal, D., and Petri, W.A., Jr (2010). Association between TNF-alpha and *Entamoeba histolytica* diarrhea. *Am. J. Trop. Med. Hyg.* 82: 620-625. doi: 10.4269/ajtmh.2010.09-0493.

Phillips, B.P., Wolfe, P.A., Rees, C.W., Gordon, H.A., Wright, W.H., and Reyniers, J.A. (1955). Studies on the ameba-bacteria relationship in amebiasis; comparative results of the intracecal inoculation of germfree, monocontaminated, and conventional guinea pigs with *Entamoeba histolytica*. *Am. J. Trop. Med. Hyg.* 4: 675-692. doi: 10. 4269/ajtmh.1955.4.675.

Picazarri, K., Luna-Arias, J.P., Carrillo, E., Orozco, E., and Rodriguez, M.A. (2005). *Entamoeba histolytica*: identification of EhGPCR-1, a novel putative G protein-coupled receptor that binds to EhRabB. *Exp. Parasitol.* 110: 253-258. doi: 10.1016/j.exppara.2005.02.014.

Prakash, V., Jackson-Akers, J.Y., and Oliver, T.I. (2020 Jan-). "Amebic Liver Abscess. 2019 Nov 7", in: *StatPearls [Internet]*. (Treasure Island, Florida, USA: StatPearls Publishing).

Rafiei, A., Ajami, A., Hajilooi, M., and Etemadi, A. (2009). Th-1/Th-2 cytokine pattern in human amoebic colitis. *Pak. J. Biol. Sci.* 12: 1376-1380. doi: 10.3923/pjbs.2009.1376.1380.

Ralston, K.S., and Petri, W.A., Jr (2011). Tissue destruction and invasion by *Entamoeba histolytica*. *Trends Parasitol.* 27: 254-263. doi: 10.1016/j.pt.2011.02.006.

Ralston, K.S., Solga, M.D., Mackey-Lawrence, N.M., Somlata, Bhattacharya, A., and Petri, W.A., Jr (2014). Trogocytosis by *Entamoeba histolytica* contributes to cell killing and tissue invasion. *Nature* 508: 526-530. doi: 10.1038/nature13242.

Ramos, E., Olivos-García, A., Nequiz, M., Saavedra, E., Tello, E., Saralegui, A., et al. (2007). *Entamoeba histolytica*: apoptosis induced in vitro by nitric oxide species. *Exp. Parasitol.* 116: 257-265. doi: 10.1016/j.exppara.2007.01.011.

Ramos, F., García, G., Valadez, A., Morán, P., González, E., Gómez, A., et al. (2005). *E. dispar* strain: analysis of polymorphism as a tool for study of geographic distribution. *Mol. Biochem. Parasitol.* 141: 175-177. doi: 10.1016/j.molbiopara.2005.02.010.

Ravdin, J.I., Abd-Alla, M.D., Welles, S.L., Reddy, S., and Jackson, T.F. (2003). Intestinal antilectin immunoglobulin A antibody response and immunity to *Entamoeba dispar* infection following cure of amebic liver abscess. *Infect. Immun.* 71: 6899-6905. doi: 10.1128/iai.71.12.6899-6905.2003.

Reed, S.L., Ember, J.A., Herdman, D.S., DiScipio, R.G., Hugli, T.E., and Gigli, I. (1995). The extracellular neutral cysteine proteinase of *Entamoeba histolytica* degrades anaphylatoxins C3a and C5a. *J. Immunol.* 155: 266-274.

Rico, G., Leandro, E., Rojas, S., Giménez, J.A., and Kretschmer, R.R. (2003). The monocyte locomotion inhibitory factor produced by *Entamoeba histolytica* inhibits induced nitric oxide production in human leukocytes. *Parasitol. Res.* 90: 264-267. doi: 10.1007/s00436-002-0780-7.

Rodríguez, M.A., Hernández, F., Santos, L., Valdez, A., and Orozco, E. (1989). *Entamoeba histolytica*: molecules involved in the target cell-parasite relationship. *Mol. Biochem. Parasitol.* 37: 87-99. doi: 10.1016/0166-6851(89)90105-9.

Rodríguez, M.A., and Orozco, E. (1986). Isolation and characterization of phagocytosis- and virulence-deficient mutants of *Entamoeba histolytica*. *J. Infect. Dis.* 154: 27-32. doi: 10.1093/infdis/154.1.27.

Rojas-López, A.E., Soldevila, G., Meza-Pérez, S., Dupont, G., Ostoa-Saloma, P., Wurbel, M.A., et al. (2012). CCR9$^+$ T cells contribute to the resolution of the inflammatory response in a mouse model of intestinal amoebiasis. *Immunobiology* 217: 795-807. doi: 10.1016/j.imbio.2012.04.005.

Rosales, C. (2020). Neutrophils at the crossroads of innate and adaptive immunity. *J. Leukoc. Biol.* 108 (1): 377-396. doi: 10.1002/JLB.4MIR0220-574RR.

Rosales, C., and Uribe-Querol, E. (2017). Phagocytosis: A fundamental process in immunity. *BioMed Res. Int.* 2017: 9042851. doi: 10.1155/2017/9042851.

Ross, A.G., Olds, G.R., Cripps, A.W., Farrar, J.J., and McManus, D.P. (2013). Enteropathogens and chronic illness in returning travelers. *N. Engl. J. Med.* 368: 1817-1825. doi: 10.1056/NEJMra1207777.

Ryan, U., Paparini, A., and Oskam, C. (2017). New technologies for detection of enteric parasites. *Trends Parasitol.* 33: 532-546. doi: 10.1016/j.pt.2017.03.005.

Sánchez-Guillén, M.C., Pérez-Fuentes, R., Salgado-Rosas, H., Ruiz-Argüelles, A., Ackers, J., Shire, A., et al. (2002). Differentiation of *Entamoeba histolytica/Entamoeba dispar* by PCR and their correlation with humoral and cellular immunity in individuals with clinical variants of amoebiasis. *Am. J. Trop. Med. Hyg.* 66: 731-737. doi: 10.4269/ajtmh.2002.66.731.

Shamsuzzaman, S.M., and Hashiguchi, Y. (2002). Thoracic amebiasis. *Clin. Chest. Med.* 23: 479-492. doi: 10.1016/S0272-5231(01)00008-9.

Sharma, M., Bhasin, D., and Vohra, H. (2008). Differential induction of immunoregulatory circuits of phagocytic cells by Gal/Gal NAc lectin from pathogenic and nonpathogenic *Entamoeba*. *J. Clin. Immunol.* 28: 542-557. doi: 10.1007/s10875-008-9184-5.

Sharma, S., Agarwal, S., Bharadwaj, R., Somlata, Bhattacharya, S., and Bhattacharya, A. (2019). Novel regulatory roles of PtdIns(4,5)P2 generating enzyme EhPIPKI in actin dynamics and phagocytosis of *Entamoeba histolytica*. *Cell. Microbiol.* 21: e13087. doi: 10.1111/cmi.13087.

Shirley, D.T., Farr, L., Watanabe, K., and Moonah, S. (2018). A review of the global burden, new diagnostics, and current therapeutics for amebiasis. *Open Forum Infect. Dis.* 5: ofy161. doi: 10.1093/ofid/ofy161.

Sim, S., Park, S.J., Yong, T.S., Im, K.I., and Shin, M.H. (2007). Involvement of beta 2-integrin in ROS-mediated neutrophil apoptosis induced by *Entamoeba histolytica. Microbes Infect.* 9: 1368-1375. doi: 10.1016/j.micinf.2007.06.013.

Singh, A., Banerjee, T., Kumar, R., and Shukla, S.K. (2019). Prevalence of cases of amebic liver abscess in a tertiary care centre in India: A study on risk factors, associated microflora and strain variation of *Entamoeba histolytica. PLos One* 14: e0214880. doi: 10.1371/journal.pone.0214880.

Singh, R.S., Walia, A.K., Kanwar, J.R., and Kennedy, J.F. (2016). Amoebiasis vaccine development: A snapshot on *E. histolytica* with emphasis on perspectives of Gal/GalNAc lectin. *Int. J. Biol. Macromol.* 91: 258-268. doi: 10.1016/j.ijbiomac.2016.05.043.

Snow, M., Chen, M., Guo, J., Atkinson, J., and Stanley, S.L., Jr (2008). Differences in complement-mediated killing of *Entamoeba histolytica* between men and women--an explanation for the increased susceptibility of men to invasive amebiasis? *Am. J. Trop. Med. Hyg.* 78: 922-923.

Somlata, Nakada-Tsukui, K., and Nozaki, T. (2017). AGC family kinase 1 participates in trogocytosis but not in phagocytosis in *Entamoeba histolytica. Nat. Commun.* 8: 101. doi: 10.1038/s41467-017-00199-y.

Sperandio, B., Fischer, N., and Sansonetti, P.J. (2015). Mucosal physical and chemical innate barriers: Lessons from microbial evasion strategies. *Semin. Immunol.* 27: 111-118. doi: 10.1016/j.smim.2015.03.011.

Stanley, S.L., Jr (2003). Amoebiasis. *Lancet* 361: 1025-1034. doi: 10.1016/S0140-6736(03)12830-9.

Teixeira, J.E., and Huston, C.D. (2008). Participation of the serine-rich *Entamoeba histolytica* protein in amebic phagocytosis of apoptotic host cells. *Infect. Immun.* 76: 959-966. doi: 10.1128/IAI.01455-0.

Thibeaux, R., Avé, P., Bernier, M., Morcelet, M., Frileux, P., Guillén, N., et al. (2014). The parasite *Entamoeba histolytica* exploits the activities of human matrix metalloproteinases to invade colonic tissue. *Nat. Commun.* 5: 5142. doi: 10.1038/ncomms6142.

Thibeaux, R., Weber, C., Hon, C.C., Dillies, M.A., Avé, P., Coppée, J.Y., et al. (2013). Identification of the virulence landscape essential for *Entamoeba histolytica* invasion of the human colon. *PLoS Pathog.* 9: e1003824. doi: 10.1371/journal.ppat.1003824.

Tillack, M., Nowak, N., Lotter, H., Bracha, R., Mirelman, D., Tannich, E., et al. (2006). Increased expression of the major cysteine proteinases by stable episomal transfection underlines the important role of EhCP5 for the pathogenicity of *Entamoeba histolytica*. *Mol. Biochem. Parasitol.* 149: 58-64. doi: 10.1016/j.molbiopara.2006.04.009.

Tran, V.Q., Herdman, D.S., Torian, B.E., and Reed, S.L. (1998). The neutral cysteine proteinase of *Entamoeba histolytica* degrades IgG and prevents its binding. *J. Infect. Dis.* 177: 508-511. doi: 10.1086/517388.

Trissl, D., Martínez-Palomo, A., de la Torre, M., de la Hoz, R., and Pérez de Suárez, E. (1978). Surface properties of *Entamoeba*: increased rates of human erythrocyte phagocytosis in pathogenic strains. *J. Exp. Med.* 148: 1137-1143. doi: 10.1084/jem.148.5.1137.

United States Centers for Disease Control and Prevention (2019). *Entamoeba histolytica Infection* [Online]. Available: https://www.cdc.gov/parasites/amebiasis/pathogen.html?CDC_AA_refVal=https%3A%2F%2Fwww.cdc.gov%2Fparasites%2Famebiasis%2Fbiology.html [Accessed April 30, 2019].

Uribe-Querol, E., and Rosales, C. (2020). Phagocytosis: Our current understading of a universal biological process. *Front. Immunol.* 11: 1066. doi: 10.3389/fimmu.2020.01066.

Vargas-Villarreal, J., Olvera-Rodríguez, A., Mata-Cárdenas, B.D., Martínez-Rogríguez, H.G., Said-Fernández, S., and Alagón-Cano, A. (1998). Isolation of an *Entamoeba histolytica* intracellular alkaline phospholipase A2. *Parasitol. Res.* 84: 310-314. doi: 10.1007/s004360050401.

Verma, A.K., Verma, R., Ahuja, V., and Paul, J. (2012). Real-time analysis of gut flora in *Entamoeba histolytica* infected patients of Northern India. *BMC Microbiol.* 12: 183. doi: 10.1186/1471-2180-12-183.

Watanabe, K., Gatanaga, H., Escueta-de Cadiz, A., Tanuma, J., Nozaki, T., and Oka, S. (2011). Amebiasis in HIV-1-infected Japanese men: clinical features and response to therapy. *PLoS Negl. Trop. Dis.* 5: e1318. doi: 10.1371/journal.pntd.0001318.

Weber, C., Blazquez, S., Marion, S., Ausseur, C., Vats, D., Krzeminski, M., et al. (2008). Bioinformatics and functional analysis of an *Entamoeba histolytica* mannosyltransferase necessary for parasite complement resistance and hepatical infection. *PLoS Negl. Trop. Dis.* 2: e165. doi: 10.1371/journal.pntd.0000165.

Winkelmann, J., Leippe, M., and Bruhn, H. (2006). A novel saposin-like protein of *Entamoeba histolytica* with membrane-fusogenic activity. *Mol. Biochem. Parasitol.* 147: 85-94. doi: 10.1016/j.molbiopara. 2006.01.010.

Wittner, M., and Rosenbaum, R.M. (1970). Role of bacteria in modifying virulence of *Entamoeba histolytica*. Studies of amebae from axenic cultures. *Am. J. Trop. Med. Hyg.* 19: 755-761. doi: 10.4269/ajtmh. 1970.19.755.

Wong-Baeza, I., Alcántara-Hernández, M., Mancilla-Herrera, I., Ramírez-Saldívar, I., Arriaga-Pizano, L., Ferat-Osorio, E., et al. (2010). The role of lipopeptidophosphoglycan in the immune response to *Entamoeba histolytica*. *J. Biomed. Biotechnol.* 2010: 254521. doi: 10.1155/ 2010/254521.

Wuerz, T., Kane, J.B., Boggild, A.K., Krajden, S., Keystone, J.S., Fuksa, M., et al. (2012). A review of amoebic liver abscess for clinicians in a nonendemic setting. *Can. J. Gastroenterol.* 26: 729-733.

Ximénez, C., Morán, P., Rojas, L., Valadez, A., Gómez, A., Ramiro, M., et al. (2011). Novelties on amoebiasis: a neglected tropical disease. *J. Glob. Infect. Dis.* 3: 166-174. doi: 10.4103/0974-777X.81695.

Yipp, B.G., Petri, B., Salina, D., Jenne, C.N., Scott, B.N., Zbytnuik, L.D., et al. (2012). Infection-induced NETosis is a dynamic process involving neutrophil multitasking in vivo. *Nat. Med.* 18: 1386-1393. doi: 10.1038/nm.2847.

Yu, Y., and Chadee, K. (1997). *Entamoeba histolytica* stimulates interleukin 8 from human colonic epithelial cells without parasite-enterocyte contact. *Gastroenterology* 112: 1536-1547.

Zhang, Z., Wang, L., Seydel, K.B., Li, E., Ankri, S., Mirelman, D., et al. (2000). *Entamoeba histolytica* cysteine proteinases with interleukin-1 beta converting enzyme (ICE) activity cause intestinal inflammation and tissue damage in amoebiasis. *Mol. Microbiol.* 37: 542-548. doi: 10.1046/j.1365-2958.2000.02037.x.

Zhou, F., Li, M., Li, X., Yang, Y., Gao, C., Jin, Q., et al. (2013). Seroprevalence of *Entamoeba histolytica* infection among Chinese men who have sex with men. *PLoS Negl. Trop. Dis.* 7: e2232. doi: 10.1371/journal.pntd.0002232.

In: *Entamoeba*
Editor: Thomas L. Johnson
ISBN: 978-1-53618-506-5

Chapter 4

PROGRAMMED CELL DEATH (PCD) IN *ENTAMOEBA HISTOLYTICA*

María Olivia Medel Flores[1], MSc, José D' Artagnan Villalba Magdaleno[2], PhD, María del Consuelo Gómez García[1], PhD, David Guillermo Pérez Ishiwara[1], PhD and Virginia Sánchez Monroy[1,*], PhD

[1]Molecular Biomedical Laboratory, ENMYH Instituto Politécnico Nacional, Ciudad de México, México

[2]Universidad la Salle, Facultad de Ingenieria Ciudad de México, México

ABSTRACT

Amoebiasis is caused by the protozoan *Entamoeba histolytica* and primarily affects developing countries. Because effective vaccines against this parasite have not been developed, the treatment of amebiasis is primarily pharmacological. However, there are reports of treatment failure and the generation of drug-resistant clones *in vitro* and *in vivo*. For

---

* Corresponding Author's E-mail: vickysm17@hotmail.com.

this reason, the study of programmed cell death (PCD) of this parasite is important since knowledge of its underlying molecular mechanism may offer novel therapeutic tools to combat the parasite. Advances in the understanding of the PCD of *E. histolytica* have demonstrated that PCD is induced by nitric oxide species, the aminoglycoside G418, hydrogen peroxide, natural compounds and host-parasite interactions. Morphological, biochemical and genetic apoptotic signals have been observed, including cell shrinkage with an increased number of vacuoles, nuclear condensation, chromatin fragmentation, preservation of the trophozoite cell-membrane integrity, externalization of phosphatidylserine (PS), overproduction of reactive oxygen species (ROS), increases in cytosolic calcium, and decreases in intracellular potassium, ATP levels and pH. The diversity of regulatory signals suggests a calpain-dependent pathway and that a caspase-dependent mechanism could be present in *E. histolytica*. Herein, we provide an overview of studies centered on the morphological, biochemical and genetic changes during PCD of *E. histolytica*.

Keywords: PCD, apoptosis like, *Entamoeba histolytica*

INTRODUCTION

Amoebiasis is caused by *Entamoeba histolytica,* and it primarily affects developing countries. It infects around 500 million people around the world, and each year it causes 50 million cases of dysentery or liver abscesses, resulting in 100,000 deaths [1]. Globally, amebiasis is the third-most common cause of death due to a parasitic infection after malaria and schistosomiasis [1]. Amebiasis infections are endemic in developing countries with temperate and tropical climates [2]. The prevalence of amebiasis varies in different countries depending on the socioeconomic conditions [2].

In industrialized countries, amebiasis is spread by unprotected sexual intercourse among gay men, immigrants, tourists traveling from areas endemic for infection, and by HIV-positive, immunocompromised individuals [3]. The global prevalence of *E. histolytica* infection in industrialized countries, such as the United States, has been estimated at approximately 4% per year despite the presence of some high-risk groups

[4]. Epidemiological studies have shown that low socioeconomic status and unsanitary conditions are significant risk factors [2]. Furthermore, young people living in developing countries are at higher risk than those in developed regions [5].

The prevalence range is 1% to 40% in Central and South America, Asia and Africa, and from 0.2% to 10.8% in industrialized countries [5]. On the other hand, the incidence of intestinal amebiasis from 1995 to 2000 was between 1,000 and 5,000 cases per 100,000 inhabitants, annually [6], with individuals under 15 years of age being the most frequently affected, with a marked increase in children between 5 and 9 years of age. However, the SUIVE/DGE/Ministry of Health/United Mexican States in 2016 reported that the most affected age groups were between 25–44 years with 39,032 records and children between 1–4 years with estimates of 39,009 cases.

In Mexico, in the period between 1995 and 2000, the incidence of amoebic liver abscess (AHA) was 10 cases per 100,000 inhabitants [7]. Hepatic amoebiasis is a serious manifestation of a disseminated infection, which occurs in less than 1% of infected individuals [8]. The most recent official data on the incidence of AHA in Mexico showed a decrease from 8.5 cases/100,000 in 1995 to 3.66 cases per 100,000 in the year 2000 [8]. Mexico, Brazil, Nicaragua and Ecuador have observed percentages of infection with *E. histolytica* from 0% to 13.8%. In Bangladesh, new *E. histolytica* infections were found in 39% of children studied for one year, among which 10% developed diarrhea and 3% dysentery [9].

The life cycle of *E. histolytica* involves an alternation of trophozoitic growth and the periodic formation of cysts [10]. This begins after ingestion of a nonmobile cyst by the human host. The mobile form of the parasite, called the trophozoite, arises after digestion of the cyst wall during passage through the small intestine. It colonizes the large intestine, where it proliferates while feeding on the resident bacterial flora. To complete the life cycle, the trophozoites develop into tetra-nucleated cysts and pass into the external environment with the host's feces. *E. histolytica* can exist as a commensal in the human intestine, or cause dysentery or extraintestinal abscesses. The proliferation of trophozoites in the lumen of the intestine occurs in a polixenic environment where it coexists with different species of anaerobic and microaerophilic bacteria. In contrast, during invasion of the intestinal epithelium or in extraintestinal lesions such as liver abscesses, amoebas proliferate in an environment free of other micro-organisms [11].

Because effective vaccines against this parasite have not been developed, the treatment of amebiasis is primarily pharmacological. However, parasites have developed molecular mechanisms that have allowed them to adapt and survive adverse conditions, avoiding the lethal effects of the drugs used in the treatment of the diseases they generate. *E. histolytica* has been reported to develop resistance to metronidazole *in vitro* under microaerophilic conditions [12, 13].

The presence of cases of drug resistance in *E. histolytica* was confirmed by surgically isolated metronidazole-resistant trophozoites from patients with liver abscesses [14]. Moreover, there are reports of treatment failure and the generation of drug-resistant clones *in vitro* and *in vivo.* A resistance mechanism that has become very important in infectious diseases caused by protozoan parasites is multidrug resistance (MDR), which is defined as the ability of cells to develop resistance to a drug that is used as a selective agent, and then they present cross-resistance to a broad spectrum of drugs that differ biochemically and structurally. In 2002, the biology of P-glycoproteins (PGPs) was studied in *E. histolytica*, and the MDR phenotype in *E. histolytica* exhibits a phenotype of drug resistance with characteristics similar to tumor cells, including cross-resistance to other drugs, an increase in drug efflux, a decrease in its accumulation in the cytoplasm and the reversal of resistance by blockers of calcium channels such as verapamil. The MDR phenotype has also been observed in clinical isolates of *E. histolytica* [15, 16].

Furthermore, a role of PGPs in the regulation of programmed cell death has been shown. Robinson et al. (1998) [17] demonstrated that transfection with the MDR gene simultaneously confers the MDR phenotype and resistance to apoptosis induced by a lack of nutrients, by the presence of chemotherapeutic drugs, by binding of TNF-α to its receptor on the surface of the cell, or by irradiation with UV light [18]. In the increased resistance scenario, the study of apoptosis-like behavior in *E. histolytica* is important because knowledge of the underlying molecular mechanism offers novel therapeutic tools to combat the parasite by promoting its own death pathways. In addition, knowledge of the PCD pathways might help in understanding the physiological implications of parasite PCD on the host-pathogen interactions and the course of the disease.

PCD

Alfred Glücksmann in 1951 reported his observations about cell death during normal vertebrate development, and in the early 1960s, John Saunders and his colleagues demonstrated the causal mechanisms of cell death in embryological experiments, which led them to propose the term "programmed cell death" [19]. However, three pathologists finally rediscovered Glücksmann´s "cell deaths" while they were investigating similar events in tumor and organ atrophy. This fact served as a cornerstone reference for the new term "apoptosis" that appeared for the first time in a biomedical journal in 1972 [20].

Apoptosis is a complex highly regulated process by which cells undergo an immunologically silent death [21]. The process has since been recognized as a distinctive mode of "programmed" cell death, which is an energy dependent process characterized by a series of distinct morphological and biochemical alterations to the cell, such as DNA fragmentation, condensation of the cytoplasm and membrane blebbing, in which a group of cysteine proteases called "caspases" are activated by genetically determined instructions and are responsible for producing the distinctive changes associated with apoptosis [22]. According to Lockshin and other researchers celebrating the first 50 years since they described cell death as "programmed", we are just beginning to understand the different mechanisms involving biochemistry and molecular genetics that have been reviewed in detail for various organisms [23].

The mechanisms of PCD are complex, involving a cascade of molecular events. The PCD cascade can be initiated via two major pathways: the extrinsic pathway or death receptor pathway, which is activated in response to transmembrane-receptors that are members of the tumor necrosis factor (TNF) receptor gene superfamily [24]; or the intrinsic mitochondrial pathways initiated by intracellular signals and involving the release of cytochrome c from the mitochondria [22]. Both pathways converge on the activation of the proteases, which are ultimately responsible for the dismantling of components of the cell, named the execution pathway. This pathway is initiated by the activation of proteases in a specific family of cysteine proteases, "the caspases", which by cleaving a set of proteins causes disassembly of the cell. The caspase-mediated proteolytic cascade represents a central point in the apoptotic response, and therefore its initiation has to be well-controlled [22, 25]. Among these regulatory factors, the death receptors, B-cell lymphoma 2 (Bcl-2) family proteins and the tumor suppressor protein p53 are of central importance [26].

As just mentioned, the process of death is the result of an interaction between initiating stimuli and factors that determine the susceptibility of the cell to activation of the final stage of PCD. The final stage is associated with distinct morphological and biochemical changes, including cell shrinkage, altered plasma membrane permeability, externalization of phosphatidylserine (PS) residues on the outer plasma membrane, loss of mitochondrial integrity, chromatin condensation, and breakdown of the DNA by Ca^{+2} and Mg^{+2} dependent endonucleases, resulting in DNA fragments of 180 to 200 base pairs [27], membrane blebbing and separation of cell fragments into apoptotic bodies [22]. Although these cell changes are clearly seen in most multicellular organisms, it is often less clear in regards to other types of cell death than may be happening on a vast scale among known organisms.

In addition, there is some evidence that PCD can occur in the absence of caspases, and other proteases such as calpains, cathepsins, and endonucleases have been described as being able to execute PCD [28]. These proteases can be activated by several cellular organelles, including mitochondria, lysosomes, and the endoplasmic reticulum, which can act independently or collaborate with each other. Although several models of caspase-independent cell death have been described, the various death routes may overlap, and several characteristics may be displayed at the same time. Therefore, we have begun to understand that death-signaling pathways are clearly interconnected, provide numerous mechanisms for cell death programs, and could explain the existence of an ancient, cell-death machinery conserved throughout the biological kingdom.

PCD in Protozoans

The presence of apoptosis-like processes in protozoans was not widely accepted when the first papers on events similar to metazoan apoptosis were published [29, 30]. However, a growing body of experimental evidence has demonstrated the existence of programmed cell death (PCD) in a great diversity of protozoans, such as *Tripanosoma sp* [29], *Leishmania sp* [31], *Plasmodium sp* [32, 33], *Toxoplasma gondii* [34], *Trichomonas vaginalis* [35], *Giardia lamblia* [36], Trichomonads [37], and *E. histolytica* [38, 39, 40].

In protozoans, the pattern of PCD events is similar to that observed in metazoan cells. Common events such as the destruction of the normal structural organization of the nucleus due to the collapse of the chromatin into electrodense masses occur, and in many cases, the breakdown of nuclear DNA into oligonucleosomal fragments also occurs. Lack of regulation of calcium homeostasis is considered to lead to significant cell death events in parasites, as most die in the presence of high calcium concentrations rather than in the absence of this ion [41], accompanied by changes in mitochondrial transmembrane potential [42], cytochrome C release [43], caspase activity [44], and PS exposition [45]. This suggests the existence of cell-death machinery conserved as part of a genetically controlled program.

PCD in parasites has been described as an important factor for clonal selection, for evasion of the immune response and for controlling the population size of microorganisms. All of these processes have importance in the context of the pathogenesis of a disease [46]. According to the above, PCD in unicellular parasites could be selectively advantageous, if some features of this process can be set in motion without necessarily resulting in death [47]. Alternatively, single-celled organisms do form colonies and in a clonal and proliferating population, not all of the cells are equally fit to survive. Thus, it is the survival of the colony as a whole, rather than the survival of each of its members, that ensures the everlasting growth of the single-celled organism [48]. Therefore, PCD could be beneficial to the parasite to maintain clonal characteristics by self-destruction of the deficient cells in the colony and thus acted to ensure the propagation of the fittest cells for the overall benefit of the parasite's survival in the host [49, 50].

Regarding the evasion of the host´s immune response by the parasites, they achieve their successful survival by depending on various mechanisms, some of them characterized by penetrating and multiplying within cells [51]. Another is the variation of antigens on their surface [52]. Some studies have shown that apoptotic parasites with PS present in the outer leaflet of their plasma membrane modulate host immunity by limiting the inflammatory response of the host [53]. Additionally, the formation of apoptotic bodies has been seen [54]. The engulfment of apoptotic bodies by phagocytes prevents cellular contents from being exposed to the immune system and it modulates responses by suppressing inflammation, modulating cell killing, and regulating immune responses [55]. Unfortunately, the adaptive immune response is not sufficient to achieve immunity, and the parasite can establish chronic infections in most individuals.

Other benefits that have been proposed for PCD of single-celled organisms is control of the population size to prevent the host's death. It is known there are correlations between parasite density and the occurrence of morphological and biochemical markers for apoptosis, and the sensing of population sizes via distinct environmental cues. Apoptosis has also been proposed as a mechanism to avoid limitations of nutrients and resources as a result of crowding [56]; or because a large number of parasites may activate immune responses that damage the parasites [57]. Then, PCD appears to regulate cell densities at least under certain conditions, and it critically affects the parasite-host interactions by facilitating a sustained parasite-host equilibrium [51].

The finding of PCD in protozoan parasites such as *Toxoplasma, Trypanosoma, Leishmania* and *Theileria* suggests that PCD is an important biological process only noticed within the last decade. The discovery and understanding of alternative death pathways will reveal new perspectives for the treatment of diseases.

PCD IN *ENTAMOEBA HISTOLYTICA*

E. histolytica is a protozoan parasite that causes amebiasis, often considered a disease of developing countries, but it is an important public health problem throughout the world. Although there have been many advances in research that may directly affect the control of amebiasis, there is still much to investigate in regard to the improvement of diagnosis and therapy. We know that PCD is an essential process in the growth and development of organisms, and according to the accumulated evidence on the cell death of *E. histolytica* with apoptosis-like features, the purposeful induction of PCD may ultimately be useful and considerably enhance our perception of the balance between health and disease.

PCD studies of *E. histolytica* have demonstrated that PCD could be induced in this parasite by nitric oxide species [39], the aminoglycoside G418 [38], and hydrogen peroxide [40]. Moreover, the natural compounds flavan-3-ol,(-)-epicatechin [58] and resveratrol [59] have also been demonstrated to be PCD inductors. The host-parasite interactions were found to involve PCD in a modified *in vivo* system [60]; in this *in vivo* system, we demonstrated that PCD played an important role in the host-parasite interactions and could thereby facilitate a sustained infection. Herein, we provide a brief overview of studies about the morphological, biochemical and genetic changes during PCD in *E. histolytica.*

Ramos et al., (2007) demonstrated that nitric oxide species (NOs) (sodium nitroprusside, sodium nitrite and sodium nitrate) were capable of inducing *in vitro* apoptosis in *E. histolytica* trophozoites. The parasites exhibited PCD characteristics such as DNA fragmentation demonstrated by electrophoretic DNA analysis and terminal deoxynucleotidyl transferase mediated dUTP nick-end labeling (TUNEL). Lower levels of ATP were also observed, suggested a high energetic demand for ATP, similar to what happens in other cells during apoptosis. There was increased proteolytic activity, and the caspase inhibitor Z-VAD-FMK and E64, a specific cysteine protease inhibitor, completely inhibited cysteine protease (CPA) activity but had no effect on the apoptosis, suggesting that NOs induced apoptosis through a caspase-independent mechanism [39].

In 2007, our group also described PCD in *E. histolytica* in response to the aminoglycoside G418 as a unique set of morphological and biochemical features. We detected morphological changes by flow cytometry and electron microscopy that included cell shrinkage with an increased number of vacuoles, nuclear condensation, chromatin fragmentation, and preservation of trophozoite cell-membrane integrity. Specific biochemical changes, such as reactive oxygen species (ROS) overproduction, an increase in cytosolic calcium, a decrease in intracellular potassium concentration and a decrease in intracellular pH were also present. Moreover, we demonstrated DNA degradation with the highly sensitive TUNEL technique, and it clearly demonstrated that E-64 blocks approximately 85% of DNA degradation, suggesting the participation of cysteine proteases in PCD induction [38].

In Mexico, alternative medicine is used to treat several diseases, including parasitic infections. Flavan-3-ol, (−)-epicatechin, an antiamoebic compound from medicinal plants *(Rubus coriifolius and Geranium mexicanum.)*, was reported as an inducer of PCD in *E. histoltyica* by Soto et al. 2010. They reported ultrastructural changes of *E. histolytica* caused by the flavan-3-ol, (−) - epicatechin, analyzed by electronic microscopy. Epicatechin induced chromatin redistribution, forming small clumps around the nuclear membrane. The trophozoites also showed a significant increase in the number of glycogen deposits and a reduction in the number and size of the vacuoles. The results showed the potential of this molecule as a possible candidate for amoebiasis chemotherapy [58].

Subsequently, using cDNA-AFLP, we searched for putative genes that could be differentially expressed during the initial phase of G418-induced PCD, using RT q-PCR; the expression patterns of the genes found were evaluated. We demonstrated that the *E. histolytica* PCD is a process that involves the expression of anti-apoptotic genes (glutaminyl-tRNA synthase, silent information regulator-2 (Sir-2), and grainins 1 and 2), in the early stages of the induction of the phenomenon followed by the expression of pro-apoptotic genes in the intermediate stages (ribosomal subunit proteins 40S and 18S and saposin-like), which give rise to effector proteins and/or execution phase specific proteins that modulate the biochemical, morphological and ultrastructure changes that mark and promote the programmed cell death. In this study, the grainins showed a positive regulatory pattern at the beginning of the PCD. Taking into account previous results in which the phenomenon was characterized by an early increase in cytosolic calcium [38], and the participation of the cysteine proteases in PCD induction, the fact that the expression of grainins increased early strongly suggested that the trophozoites activated molecular mechanisms trying to recapture free cytosolic calcium, and probably the PCD induction involved calcium-dependent proteases [61].

Ghosh et al., also in 2010, demonstrated morphological characteristics of PCD in the parasites when they were exposed to hydrogen peroxide. They underwent DNA fragmentation as shown by TUNEL assays, externalization of PS, and an increase in the level of reactive endogenous oxygen species using flow cytometry. Although the parasite genome does not have canonical caspases, it still encodes a large number of cysteine proteases that can assume the apoptotic function of caspases. Furthermore, they demonstrated that E-64 does not have any effect on the hydrogen peroxide mediated cell death in *E. histolytica*, suggesting an independent pathway of cysteine proteases [40].

In parallel, another study of PCD in *E. histolytica* induced by hydrogen peroxide detected morphological and biochemical PCD characteristics. The morphological characteristics included DNA fragmentation, increased vacuolization, nuclear condensation, and cell rounding with preservation of the membrane integrity. Biochemical alteration of ion fluxes is also a key feature in PCD, and trophozoites exposed to hydrogen peroxide showed overproduction of reactive oxygen species, increased cytosolic Ca^{2+}, and decreased intracellular pH. Furthermore, it was found that PS was exposed on the plasma membrane of the trophozoites. The trophozoites also showed increased calpain activity, indicating the involvement of calpain-dependent Ca^{2+}-cysteine proteases in *E. histolytica* PCD. These observations clearly demonstrate a calpain-dependent pathway for oxidative apoptosis-like death of the *E. histolytica* trophozoites [62].

Our research group also studied the PCD of *E. histolytica* in a modified *in vivo* system. To study the *in vivo* events related to the inflammation and host-parasite interactions during amebiasis, we utilized a designed host-pathogen interaction model in which cellulose membrane dialysis bags containing *E. histolytica* trophozoites were surgically placed inside the peritoneal cavity of a hamster and left to interact with the host's exudate components. This model was developed to determine if cellular immunity contributes to the damage of *E. histolytica* trophozoites by releasing molecules that induce amebic-PCD. Amoeba collected from the bags were examined by TUNEL assays and electron microscopy, the results showed nuclear condensation and DNA fragmentation after exposure to peritoneal exudates, which were composed of neutrophils and macrophages. Our results also demonstrated the presence of nitric oxide (NO) in the supernatant contained in the bag, which suggests that the production of nitric oxide by inflammatory cells could be involved in the PCD of trophozoites. These results demonstrated that apoptotic parasites play a role in host-parasite interactions, thereby facilitating a sustained infection [60].

In 2013, our group studied the modulation of PCD induced by G418 through *E. histolytica* P-glycoprotein (EhPgp). In mammals, intracellular acidification serves as a global switch for inactivating cellular processes, and it initiates molecular mechanisms implicated in the destruction of the genome. In contrast, intracellular alkalinization produced by P-glycoprotein overexpression in multidrug-resistant cells has been related to apoptosis resistance.

In this study, we showed that antisense inhibition of PGP expression produced a synchronous death of trophozoites and the enhancement of biochemical and morphological characteristics of PCD induced by G418. The nucleus was contracted, and the nuclear membrane was disrupted. Moreover, the chromatin was extensively fragmented. The Ca^{2+} concentration was increased, while the intracellular pH was acidified. In contrast, PGP overexpression prevented intracellular acidification and circumvented the apoptotic effect of G418. In conclusion, PGP blocked the global switch that inactivates cellular processes and the molecular mechanisms implicated in the destruction of the genome represent a role for PCD resistance. Our results were the first evidence in parasites and other unicellular organisms of a specific molecule participating in the regulation of PCD pathways [63].

Subsequently, considering the report of Nandy et al. 2010 [62], which demonstrated a calpain-dependent pathway for apoptosis-like death that was induced by H_2O_2, and our previous reports of PCD induced by G418, in which calcium release was concomitant with *de novo* specific protein synthesis, we identified the participation of calpain-like proteases in PCD induced by G418. We evaluated the expression and activity of this calpain-like effect by RT-qPCR, western blot assays, and an enzymatic method during the induction of PCD by G418. In addition, using cell viability and TUNEL assays, we also demonstrated that the Z-Leu-Leu-Leu-al calpain inhibitor reduced the rate of cell death. These results suggest that calpain-like proteases could be involved in the executory phase of PCD [64].

Pais-Morales et al., 2016, demonstrated PCD in *E. histolytica* trophozoites after exposure to resveratrol (trans-3,4',5- trihydroxystilbene) [59], a nontoxic to humans natural compound with antioxidant properties present in red wine, grapes nuts and medicinal plants [65]. Resveratrol generated the production of ROS, lipid peroxidation, PS externalization, DNA fragmentation, an increase of intracellular Ca^{2+} concentration, activated calpain, and decreased superoxide dismutase activity, indicating a PCD event. Moreover, they demonstrated that cytopathic activity, phagocytosis, encystment, and virulence *in vivo* were decreased by preincubation of the trophozoites with resveratrol, showing that resveratrol decreased the virulence of trophozoites *in vitro*. Interestingly, after inoculation with virulent trophozoites, hamsters treated with this drug did not develop abscesses or only developed very small abscesses. These results suggest that resveratrol could be an alternative to avoid amoebic hepatic abscesses.

In 2018, our group considered our previous report, which described that calpain-like proteases could be involved in the executory phase of PCD induced by G418, demonstrating that calpain-like protein plays an important role in the execution phase of *E. histolytica* PCD. In this study, we modeled a hypothetical 3D structure of the calpain-like protein of *E. histolytica* and produced specific antibodies against it. Using western blot (WB) assays and confocal microscopy analyses, we demonstrated that the expression of calpain-like protein gradually increased during PCD induction, localizing the protein in the cytoplasm and near the nucleus. Knockdown of the calpain-like gene by a specific small interference RNA sequence reduced the cell death rate by 65%. These results support the hypothesis that calpain-like protein plays an important role in the execution phase of *E. histolytica* PCD. In addition, a hypothetical interactome of the calpain-like protein suggests that other proteins, including some with calcium-binding domains, also participate in the PCD pathway of this parasite [66].

Recently, Valle-Solis et al., [67] identified a calcium/cation exchanger that is a protein that regulates the calcium flux in this parasite. They analyzed its possible role in some cellular processes triggered by calcium flux, such as the PCD and virulence. First, by searching for putative calcium/cation exchangers in the genome database of *E. histolytica*, they identified a protein belonging to the CCX family (EhCCX). After they generated a specific antibody against EhCCX, they showed that this protein was expressed at higher levels in *E. histolytica* than in the nonpathogenic amoeba *E. dispar*. In addition, the expression of EhCCX was increased in trophozoites incubated with hydrogen peroxide (an inducer of PCD in *E. histolytica* that has been previously described). This

E. histolytica exchanger was localized in the plasma membrane and in the membranes of some cytoplasmic vesicles. The parasites that overexpressed this exchanger contained higher cytosolic calcium levels than the control, but the extrusion of calcium after the addition of hydrogen peroxide was more efficient in the EhCCX-overexpressing trophozoites; consequently, the PCD was delayed in these parasites. Interestingly, the overexpression of EhCCX increased the *in vitro* virulence demonstrated by the properties of the trophozoites, such as phagocytosis, destruction of mammalian cell monolayers and migration. These results suggest that EhCCX plays important roles in the PCD and in the *in vitro* virulence of *E. histolytica* [67].

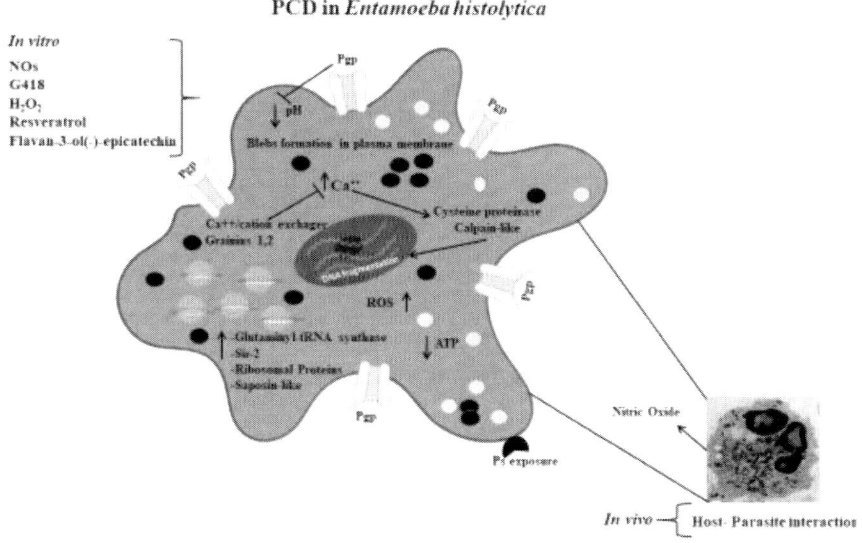

Figure 1. PCD signals in *E. histolytica.*

Studies conducted over the past 15 years have revealed a complex of death machinery that regulates PCD. The death machinery of *E. histolytica* can be activated by diverse stimuli and has as its central components a group of specific proteolytic enzymes.

Table 1. PCD Signals in *Entamoeba histolytica*

PCD Signal	Inducer of PCD	Reference
Morphological PCD Signals		
Cell Shrinkage	NOs G418 Flavan-3-ol,(−)-epicatechin H_2O_2 Resveratrol	[39] [38, 63, 66] [58] [40, 62] [59]
Nuclear condensation	G418 H_2O_2 Host-parasite interaction	[38, 63] [62] [60]
Increased number of vacuoles	G418 H_2O_2	[38] [62]
Decreased number of vacuoles	Flavan-3-ol, (−)-epicatechin	[58]
Chromatin redistribution	Flavan-3-ol, (−)-epicatechin	[58]
DNA fragmentation	NOs G418 H_2O_2	[39] [38, 60, 63, 64, 66] [40, 59, 62]
PS exposure	H_2O_2 Resveratrol	[40, 62] [59]
Biochemical PCD signals		
Activity of cysteine proteases	G418	[38]
Increased intracellular Ca^{2+}	NOs H_2O_2 G418	[39] [40, 59, 62] [38, 63]
Decreased intracellular K^+	G418	[38]
Increased ROS	G418 H_2O_2 Resveratrol	[38] [40, 5] [2]
Decreased intracellular pH	G418	[38, 63]
Decreased ATP levels	NOs	[39]
Lipid peroxidation	Resveratrol	[59]
Genetic signals in the PCD regulation		
Anti-apoptotic signals: Glutaminyl-tRNA synthase Silent Information Regulator-2 (Sir-2) Grainins 1 and 2 **Apoptotic signals:** Ribosomal Subunit Proteins Saposin-like	G418	[61]
EhPgp inhibition induced PCD	G418	[63]
Overexpression Calpain-like induced PCD	H_2O_2 G418	[62, 59, 67] [64, 66]
Overexpression Calcium/Cation Exchanger decreased PCD	H_20_2	[67]

Reports of these parameters can provide valuable data on the molecular mechanisms of PCD and for the identification of appropriate targets for chemotherapeutic treatment. Much remains to be learned about the complexity of PCD, because it appears to be a mechanism important to the survival of almost all known organisms.

Table 1 and Figure 1 summarize the evidence of PCD signals in *E. histolytica.*

CONCLUSION

PCD is a vital process that is intimately involved in the control of cellular homeostasis throughout the life span of organisms. Moreover, this process of death is clearly different among the organisms in accordance with the mode of death described in the literature. It appears to be a consequence of extreme perturbation of the cellular microenvironment; otherwise, the organism can successfully resolve the stress exerted. The adaptation of parasitic protozoa to their hosts involves modulation of cell death in order to facilitate parasite survival in a hostile environment. Thus, the molecular dissection of the PCD indicates a blueprint for apoptosis that is conserved during evolution and will provide new opportunities for the treatment of various human diseases.

Advances in the study of PCD in *E. histolytica* have demonstrated that PCD could be induced in this parasite by nitric oxide species, the aminoglycoside G418, and hydrogen peroxide. Moreover, the natural compounds flavan-3-ol,(-)-epicatechin and resveratrol have also been demonstrated to be PCD inducers. Finally, host-parasite interactions have also been shown to induce PCD in a modified *in vivo* system; in this *in vivo* system, we demonstrated that PCD plays an important role in the host-parasite interactions and could thereby facilitate a sustained infection [60]. In this sense, *E. histolytica* plays an important function in the adhesion, killing or phagocytosis of target cells. During infection of colonic tissues, amoebic trophozoites are able to kill host cells via apoptosis or necrosis and trigger acute inflammatory responses. Lee et al., [68] showed in an *in vitro* assay that amoeba-induced apoptotic cell death of Jurkat T cells demonstrated that target cell killing required lectin mediated contact, calcium influx and tyrosine dephosphorylation activation. Some results suggest that calpain plays an important role in this protozoan parasite for inducing the degradation of important molecules like NK-kB and STAT [69], altering in this way a prototypical proinflammatory signaling pathway, which ultimately accelerates cell death. Recently, López-Rosas et al., [70] showed the impact of *E. histolytica* on host microRNAs as a prominent class of negative regulators of gene expression, and it may provide new insights into the relationship between this parasite and human cells. These observations support a model where cytotoxic effectors are secreted in a regulated, contact-dependent manner. However, the signaling pathways involved in host cell death induced by *E. histolytica* have not yet been fully defined; some are amoebapores, saposin-like proteins, cysteine proteases, acid vesicle components and other factors that contribute to tissue invasion.

It is generally accepted that PCD signaling is constitutively expressed in most if not all cells; the reported data suggest the existence of PCD in *E. histolytica* initiated by a variety of stress stimuli, which active processes of cellular self-destruction with distinctive morphological and biochemical features described below. The apoptotic signals evidenced in the PCD of this parasite by different inducers include cell shrinkage with an increased number of vacuoles, nuclear condensation, chromatin fragmentation, and, preservation of trophozoite cell-membrane integrity, externalization of PS, overproduction of reactive oxygen species (ROS), increase in cytosolic calcium, decrease in intracellular potassium concentration, and decrease in intracellular pH, and it also undergoes a period of blebbing that was visualized in time-lapse photography.

A diversity of signals have also been noted in response to different inducers, and overexpression of calcium/cation exchanger, glutaminyl-tRNA synthase, silent information regulator-2 (Sir-2), and grainins 1 and 2 have all been shown to participate as anti-apoptotic signals. Overexpression of ribosomal subunit proteins 40s, saposin-like, p-glycoprotein (EhPgp) and calpains have been shown to trigger pro-apoptotic signals. Together, these signals use a calpain-dependent pathway, which suggest that a caspase-dependent mechanism could be present in *E. histolytica.* Although the parasite genome is devoid of any of the mammalian caspase homologs, it encodes several cysteine proteases that can assume the apoptotic function of caspases [40]. Moreover, *E. histolytica* trophozoites do not possess canonical mitochondria, but do have cytoplasmic organelles with DNA [71] and mitochondrial-like enzymes [72, 73]. The mitosomes found in *E. histolytica* trophozoites hints at a biological significance and a reason for these organelles being in this parasite [71]. On the other hand, Santos et al., [74] reported they localize to vacuolar compartments/Golgi-like structures that allow them to respond similar to a eukaryotic cell.

In summary, initial database searches indicated that the *E. histolytica* genome lacks genes with obvious homology to many critical components of metazoan apoptotic death machinery, such as the Bcl-2 family of proteins or caspases that play an important role in the control of apoptosis. That led to the conclusion that PCD does not exist in this parasite. However, new evidence has accumulated during the last 15 years, implying that amoebic trophozoites can undergo PCD and many components of cell death. *E. histolytica* and metazoan apoptosis may be evolutionarily conserved and provide a basis for understanding programmed cell death in more complex organisms. In addition, we demonstrated that the induction of PCD is correlated with the level of intracellular Ca^{2+}, and we proposed that a calpain-like cysteine protease that is activated by a calcium gradient takes part in cell death. Calpain has been present throughout evolution and is found in almost all eukaryotes. Significant morphological and biochemical changes during the process of PCD occur in *E. histolytica* as a reference for this complex phenomenon. We detailed the morphology and biochemical features of PCD in this important parasite with specific and selective techniques, which are optimal to target hallmark apoptotic features, such as specific fluorescent dyes, confocal and electron microscopy, TUNEL assays, DNA fragmentation analysis by gel electrophoresis, flow cytometers and molecular tools. It should be clear that PCD is a complex phenomenon. It is unlikely that this complexity is merely a reflection of the exclusively *in vitro* approaches. Recent studies of the various mechanisms used by parasites have shed light on the structural, biochemical and genetics responsible for regulating cell death, and of interest are questions about apoptosis in immune, inflammatory and stress responses.

REFERENCES

[1] *World Health Organization.* 1997. Bull W.H.O. 97-100.

[2] Tanyuksel M and Petri Jr W.A. 2003. "Laboratory diagnosis of amebiasis." *Clinical Microbiology Reviews* 16: 713-29. doi: 10.1128/CMR.16.4.

[3] Watanabe Koji, Gatanaga Hiroyuki, Escueta-de Cadiz Aleyla, Tanuma Junko, Nozaki Tomoyoshi and Oka Shinichi. 2011. "Amebiasis in HIV-1-infected Japanese men: clinical features and response to therapy." *PLoS Neglected Tropical Diseases* 5 e1318. doi:10.1371/journal.pntd.0001318.

[4] Gunther Janelle, Shafir Shira, Bristow Benjamin and Sorvillo Frank. 2011. "Short report: Amebiasis-related mortality among United States residents, 1990-2007." *The American journal of Tropical Medicine and Hygiene* 85: 1038-40. doi:10.4269/ajtmh.2011.11-0288.

[5] Braga Lucia L, Mendoca Yacy, Paiva Clece A, Sales Andrea, Cavalcante Andre LM, Mann Barbara J. 1998. "Seropositivity for and Intestinal Colonization with *Entamoeba histolytica* and *Entamoeba dispar* in Individuals in Northeastern Brazil". *Journal of Clinical Microbiology* 36: 3044-45.

[6] Ximenez C, Moran P, Rojas L., Valadez A. and Gomez, A. (2009). "Reassessment of the epidemiology of amebiasis: state of the art". *Infection Genetics and Evolution.* 9: 1023-32. doi: 10.1016/j. meegid.2009.06.008.

[7] González Vázquez MC, Rosales-Encina JL, Carabarin-Lima A. 2012. "Of Amebiasis and Amebiasis: *Entamoeba histolytica*". *Elementos* 87: 13-18.

[8] Rodríguez R, Carbajal L, Perea A, Pérez L, Lizarraga S and Campos T. 2010. "Amebic Liver Abscess complicated with intra-abdominal and thoracic rupture." *Revista de Enfermedades Infecciosas en Pediatría, XXIV* 64-8.

[9] Haque R, Mondal D, Duggal P, Kabir M, Roy S, Far BM. 2006. "*Entamoeba histolytica* infection in children and protection from subsequent amebiasis". *Infection and Immunity* 74: 904-9. doi:10.1128/IAI.74.2.904–909.2006.

[10] Haque Rashidul, Huston Christopher D, Hugher Molly, Houpt Eric, Petri Jr William A. 2003. "Amebiasis". *The New England Journal of Medicine* 348:1565-73. doi: 10.1056/NEJMra022710.

[11] Mukherjee Chandrama, Graham Clartk C and Lohia Anuradha. 2008. "*Entamoeba* shows reversible variation in Ploidy under different growth conditions and between life cycle phases". *PLOS Neglected Tropical Diseases* 8:e281. doi:10.1371/journal.pntd.0000281.

[12] Samarawickrema Nirma A, Brown David M, Upcroft Jaqueline A. 1997. Thammapalerd Nitaya and Upcroft P. "Involvement of superoxide dismutase and pyruvate: ferredoxina oxidoreductase in mechanism of metronizadole resistance in *Entamoeba histolytica*". *Journal of Antimicrobial Chemotherapy* 40: 883-40.

[13] Wassmann Claudia, Hellberg Andrea, Tannich Egbert and Bruchhaus Iris. 1999. "Metronidazole Resistance in the protozoan Parasite *Entamoeba histolytica* is associated with increased Iron-containing Superoxide Dismutase and Peroxiredoxin and decreased expression of Ferredoxin 1 and Favin Reductase". *The Journal of Biological Chemistry* 274: 26051-56.

[14] Hanna RM, Dahniya MH, Badr SS and El-Betagy A. 2000. "Percutaneous catheter drainage in drug-resistant amoebic liver abscess". *Tropical Medicine and International Health* 5: 578-81. doi10.1046/j.1365-3156.2000.00586.x.

[15] Bañuelos Cecilia, Orozco Esther, Gómez Consuelo, González Arturo, Medel Olivia, Mendoza Leobardo and Pérez D. Guillermo. 2002. "Cellular Location and Function of the P-Glycoproteins (EhPgps) in *Entamoeba histolytica* Multidrug-Resistant Trophozoites". *Microbial Drug Resistance* 8: 291-300. doi: 10.1089/10766290260469552.

[16] Bansal Devendra, Malla Nancy and Mahajan R.C.2006. "Drug resistance in amoebiasis". *Journal of Medicine Research* 123: 115-118.

[17] Robinson Laura, Roberts Wendy K, Ling Tao Tao, Lamming Dudley, Sternberg Stephen S and Roepe Paul D. 1997. "Human MDR 1 Protein Overexpression Delays the Apoptotic Cascade in Chinese Hamster Ovary Fibroblasts". *Biochemestry* 37: 11169-78. doi: 10.1021/bi9627830.

[18] Johnston RW and Smyth MJ. 2000. "Multiple Physiological Functions for Multidrug Transporter P-glycoprotein". *Trends Biochemical Sciences* 25: 1-6, doi: 10.1016/s0968-0004(99)01493-0.

[19] Saunders John W. 1966. "Death in Embryonic Systems". *Science* 3749:604-12. doi:10.1126/science.154.3749.604.

[20] Kerr, JF, Wyllie H, Currie AR. 1972. "Apoptosis: A basic Biological Phenomenon with Wide-Ranging Implications in Tissue Kinetics." *British Journal Cancer* 26:239-57. doi:10.1038/bjc.1972.33.

[21] Carmen John C and Sinai Anthony P. 2007. "Suicide prevention: disruption of apoptotic pathways by protozoan parasites." *Molecular Microbiology* 64:904-16. doi:10.1111/j.1365-2958.2007.05714.x.

[22] Elmore Susan. 2007. "Apoptosis: A Review of Programmed Cell Death." *Toxicologic Pathology* 35:495–516. doi:10.1080/01926230701320337.

[23] Lockshin RA. 2016. "Programmed cell death 50 (and beyond)." *Cell Death Differentiation* 23:10-7doi: 10.1038/cdd.2015.126.

[24] Locksley RM, Killeen N and Lenardo MJ. 2001. "The TNF and TNF Receptor Superfamilies: Integrating Mammalian Biology". *Cell* 104: 487-501. doi: 10.1016/s0092-8674(01)00237-9.

[25] Green DR, Fabien Llambi. 2015. "Cell death signaling." Cold Spring Harbor Perspectives in Biology 7, doi:10.1101/cshperspect. a006080.

[26] Assuncao GC and Rafael Linden. 2004. "Programmed Cell Death Apoptosis and Alternative Deathstyles." *European Journal of Biochemistry* 271:1638-50.doi:10.1111/j.1432-1033.2004.04084.x.

[27] Bortner CD, Oldenburg NB, and Cidlowski JA. 1995. "The role of DNA fragmentation in apoptosis". *Trends in Cell Biology* 5:21–6.doi:10.1016/s0962-8924(00)88932-1.

[28] Broker Linda E, Kruyt Frank AE and Giaccone Guiseppe. 2005. "Cell Death Independent of Caspase: A review." *Clinical Cancer Research* 11:3155-62. doi: 10.1158/1078-0432.CCR-04-2223.

[29] Ameisen JC, Idziorek T, Billaut-Mulot O, Loyens M, Tissier JP, Potentier A and Quaissi A. 1995. "Apoptosis in a Unicellular Eukaryote (*Trypanosoma Cruzi*): Implications for the Evolutionary Origin and Role of Programmed Cell Death in the Control of Cell Proliferation, Differentiation and Survival." *Cell Death Differentiation* 4: 285-300.

[30] Jiménez-Ruiz Antonio, Alzate Juan Fernando, Macleod Ewan Thomas, Lüder Carsten Günter Kurt, Fasel Nicolas and Hurd Hilary. 2010. "Apoptotic markers in protozoan parasites." *Parasites &Vectors* 3: 1-15.

[31] Basmaciyan Louise and Casanova Magali. 2019. "Cell Death in Leishmania" *Parasite* 26:71. doi:10.1051/parasite/2019071.

[32] Al-Olayan, Williams and Hurd H. 2002. "Apoptosis in the malaria protozoan, *Plasmodium berghei*: a possible mechanism for limiting intensity of infection in the mosquito." *International Journal for Parasitology* 32: 1133-1143.

[33] Chang JH, Kotturi SR, Chong AGL, Lear M J and Tan K SW. 2010. "A Programmed Cell Death Pathway in the Malaria Parasite Plasmodium Falciparum Has General Features of Mammalian Apoptosis but Is Mediated by Clan CA Cysteine Proteases." *Cell Death & Disease* 1 e26:1-13. doi:10.1038/cddis.2010.2.

[34] Peng Lin, Lin Y, Jiang and Zhang T. 2003. "Exogenous Nitric Oxide Induces Apoptosis in *Toxoplasma Gondii* Tachyzoites via a Calcium Signal Transduction Pathway." *Parasitology* 126:541-50.

[35] Chose, Oliver, Nöel Christophe, Gerbod Delphine, Brenner Catherine, Viscogliosi Eric and Roseto Alberto. 2002. "A Form of Cell Death with Some Features Resembling Apoptosis in the Amitochondrial Unicellular Organism Trichomonas Vaginalis." *Experimental Cell Research* 276:32-9. doi: 10.1006/excr.2002.5496.

[36] Ghosh E, Ghosh A, Ghosh AN, Nozaki and Ganguly S. 2009. "Oxidative stress-induced cell cycle blockage and a protease-independent programmed cell death in microaerophilic *Giardia lamblia.*" *Drug Dosing Development and Therapy* 3: 103-110, doi: 10.2147/dddt.s5270.

[37] Benchimol Marlene, Rosa Ivone de Andrade, Fontes Reginaldo da Silva and Dias Burla Angelo José. 2008. "Trichomonas Adhere and Phagocytose Sperm Cells: Adhesion Seems to Be a Prominent Stage During Interaction." *Parasitology Research* 102:597-604. doi:10.1007/s00436-007-0793-3.

[38] Villalba J D'Artagnan, Gómez Consuelo, Medel Olivia, Sánchez Virginia, Carrero Julio C, Shibayama Mineko and Perez D Guillermo. 2007. "Programmed Cell Death in *Entamoeba histolytica* Induced by the Aminoglycoside G418." *Microbiology* 153: 3852-63. doi:10.1099/mic.0.2007/008599-0.

[39] Ramos Espiridión, Olivos-García Alfonso, Nequiz Mario, Saavedra Emma, Tello Eusebio, Saralegui Andrés, Montfort Irmgard and Perez-Tamayo Ruy. 2007. "*Entamoeba histolytica*: Apoptosis Induced in vitro by Nitric Oxide Species." *Experimental Parasitology* 116:257.65. doi:10.1016/j.exppara.2007.01.011.

[40] Ghosh, Anupama Sardar, Dutta Suman and Raha Sanghamitra Raha. 2010. "Hydrogen Peroxide-Induced Apoptosis-Like Cell Death in *Entamoeba histolytica.*" *Parasitology International* 59:166-72. doi:10.1016/j.parint.2010.01.001.

[41] Fernández-Presas Ana M. 2000. “Apoptosis in protozoans and in the host cell induced by protozoa.” *Revista Mexicana de Patología Clínica* 47:84-93.

[42] Irigoín F, Inada NM, Fernandes MP, Piacenza L, Gadelha FR, Vercesi AE, Radi R. (2009). “Mitochondrial Calcium Overload triggers Complement-Dependent Superoxide-Mediated Programmed Cell Death in Trypanosoma Cruzi.” *The Biochemical Journal* 418:595-604. doi: 10.1042/BJ20081981.

[43] Koutsogiannis Z, MacLeod ET, Maciver SK. 2019. “G418 Induces Programmed Cell Death in Acanthamoeba Through the Elevation of Intracellular Calcium and Cytochrome C Translocation” Parasitology Research. 118:641-51. doi: 10.1007/s00436-018-6192-0.

[44] Balakrishnan DD, Kumar SG. 2014. “Higher caspase-like activity in symptomatic isolates of Blastocystis spp.” *Parasites & Vectors.* 7:219. doi: 10.1186/1756-3305-7-219.

[45] Cárdenas-Zúñiga R, Silva-Olivares A, Villalba-Magdaleno JA, Sánchez-Monroy V, Serrano-Luna J, Shibayama M. 2017. Amphotericin B Induces Apoptosis-Like Programmed Cell Death in *Naegleria Fowleri* and *Naegleria Gruberi*. *Microbiology.* 163:940-949. doi: 10.1099/mic.0.000500.

[46] Bruchhaus Iris, Roeder Thomas, Rennenberg Annika and Heussler Volker T. 2007. “Protozoan Parasites: Programmed Cell Death as a Mechanism of Parasitism.” *Trends Parasitology.* 23: 376-83. doi:10.1016/j.pt.2007.06.004.

[47] DosReis GA and Barcinski, MA. 2001. “Apoptosis and Parasitism: From the Parasite to the Host Immune Response.” *Advances Parasitology* 49: 133-61. doi:10.1016/s0065-308x(01)49039-7.

[48] Ameisen, JC. 2002. “On the origin, evolution, and nature of programmed cell death: a timeline of four billion years.” *Cell Death Differentiation* 9: 367-393. doi:10.1038/ sj/cdd/4400950.

[49] Moreira Maria Elisabete C, Del Portillo Hernando A, Milder Regina V Milder, Balanco Jose Mario F and Barcinski Marcello.A. 1996. “Heat Shock Induction of Apoptosis in Promastigotes of the Unicellular Organism Leishmania (Leishmania) Amazonensis.”

Journal Cell Physiology 167: 305-13. doi:10.1002/(SICI)1097-4652(199605)167:2<305:AID-JCP15>3.0.CO;2-6.

[50] Vickerman K. 1985. “Developmental Cycles and Biology of Pathogenic Trypanosomes.” *British Medical Bulletin* 41: 105-14. doi:10.1093/oxfordjournals.bmb.a072036.

[51] Lüder Carsten GK, Campo-Salinas Jenny, Gonzalez-Rey Elena and Van Zandbergen Ger van. 2010. “Impact of Protozoan Cell Death on Parasite-Host Interactions and Pathogenesis.” *Parasites & Vectors* 2: 1-11. doi:10.1186/1756-3305-3-116.

[52] Zambrano-Villa S, Rosales-Borjas D, Carrero, Julio C, Ortiz-Ortiz L. 2002. How protozoan parasites evade the immune response. *Trends in Parasitology* 18:272-278 doi: 10.1016/S1471-4922(02)02289-4.

[53] Wanderley JL, Moreira ME, Benjamin A, Bonomo AC and Barcinski MA. 2006. “Mimicry of Apoptotic Cells by Exposing Phosphatidylserine Participates in the Establishment of Amastigotes of Leishmania (L) Amazonensis in Mammalian Hosts.” *The Journal Immunology* 176:1834-9. doi: 10.4049/jimmunol.176.3.1834.

[54] Nasirudeen AM, Hian YE, Singh M and Tan KS. 2004. “Metronidazole induces programmed cell death in the protozoan parasite *Blastocystis hominis.” Microbiology* 150:33–43. doi:10.1099/mic.0.26496-0.

[55] Savil J and Fadok V. 2000. “Corpse clearance defines the meaning of cell death.” *Nature* 407:784–788. doi:10.1038/35037722.

[56] Pollitt LC, Colegrave N, Khan S, Sajid M and Reece S E. 2010. “Investigating the evolution of apoptosis in malaria parasites: the importance of ecology.” *Parasites &Vectors* 3: 105. doi:10.1186/1756-3305-3-105.

[57] Hurd H, Grant KM and Arambage SC. 2006. “Apoptosis-like death as a feature of malaria infections in mosquitoes.” *Parasitology* 132: S33-47. doi:10.1017/S0031182006000849.

[58] Soto J, Gómez C, Calzada F, Ramírez ME. 2010. “Ultrastructural changes on *Entamoeba histolytica* HM1-IMSS caused by the flavan-3-ol, (-)-epicatechin.” *Planta Medical* 76(6):611-612.

[59] Pais-Morales Jonnatan, Betanzos Abigail, Garcia-Rivera Guillermina, Chávez-Munguia Bibiana, Shibayama Mineko and Orozco Esther. 2016. "Resveratrol induces apoptosis-Like Death and Prevents *In vitro* and *In vivo* Virulance of *Entamoeba histolytica*." *PLOS ONE* 1:e0146287. doi:10.1371/journal.pone.0146287.

[60] Villalba Magdaleno Jose D´Artagnan, Pérez Ishiwara Guillermo, Serrano-Luna Jesús, Tsutsumi Victor and Shibayama Mineko. 2011. "*In vivo* programmed cell death of *Entamoeba histolytica* trophozoites in a hamster model of amoebic liver abscess." *Microbiology* 157: 1489-99. doi:10.1099/mic.0.047183-0.

[61] Sánchez Monroy Virginia, Medel Flores Ma. Olivia, Villalba-Magdaleno Jose D´Artagnan, Gómez García Consuelo and Pérez Ishiwara David Guillermo. 2010. "*Entamoeba histolytica*: Differential gene expression during programmed cell death and identification of early pro- and anti-apoptotic signals." *Experimental Parasitology* 126: 497-505. doi.org/10.1016/j.exppara.2010.05.027.

[62] Nandi Nilay, Sen Abhik, Banerjee Rajdeep, Kumar Sudeep, Kumar Vikash, Ghosh Amar Nath and Das Pradeep. 2010. "Hydrogen peroxide induces apoptosis-like death in *Entamoeba histolytica* trophozoites." *Microbiology* 156:1926-41. doi:10.1099/mic.0.034066-0.

[63] Medel Flores Olivia, Gómez García Consuelo, Sánchez Monroy Virginia, Villalba Magdaleno José D´Artagnan, Nader García Elvira and Pérez Ishiwara D. Guillermo. 2013. "*Entamoeba histolytica* P-glycoprotein (EhPgp) inhibition, induce trophozoite acidification and enhance programmed cell death." *Experimental Parasitology* 135: 532-40. doi:org/10.1016/j.exppara.2013.08.017.

[64] Sánchez Monroy Virginia, Medel Flores Olivia, Gómez García Consuelo, Chávez Maya Yesenia, Domínguez Fernández Tania and Pérez Ishiwara D. Guillermo. 2015. "Calpain-like: A Ca^{+2} dependent cystein protease in *Entamoeba histolytica* cell death." *Experimental Parasitology* 159: 245-251. doi.org/10.1016/j.exppara.2015.10.005.

[65] Park EJ, Pezzuto JM. (2015). The pharcology of resveratrol in animals and humans. *Biochemica et Biophysica Acta (BBA)-*

Molecular Basis of Disease 1852:1071-113 doi:10.1016/j. bbadis.2015.01.014.

[66] Domínguez Fernández Tania, Rodríguez Mario Alberto, Sánchez Monroy Virginia, Gómez García Consuelo, Medel Olivia, Pérez Ishiwara David Guillermo. "A Calpain-Like Protein is involved in the execution phase of Programmed Cell Death of *Entamoeba histolytica.*" *Frontiers in Cellular and Infection Microbiology* 8: 339. doi.org/10.3389/fcimb.2018.00339.

[67] Valle-Solis Martha, Bolaños Jeni, Orozco Esther, Huerta Miriam, Garcia-Rivera Guillermina, Salas-Casas Andrés, Chávez-Munguia Bibiana and Rodriguez Mario A. 2018. "A Calcium/Cation Exchanger Participates in the Programmed Cell Death and in vitro Virulance of *Entamoeba histolytica*." *Frontiers in Cellular and Infection Microbiology* 8:342. doi.org/10.3389/fcimb.2018.00342.

[68] Lee YA, Kim KA, Min A and Shin MH. 2014. "Amoebic PI3K and PKC is required for Jurkat T cell death induced by *Entamoeba histolytica". The Korean Journal Parasitology* 54:355-65. doi: https://doi.org/10.3347/kjp.2014.52.4.355.

[69] Kim KA, Min A, Lee YA, Shin MH. 2014. "Degradation of the transcription factors NF-kB, STAT3, and STAT5 is involved in *Entamoeba histolytica* cell death in Caco-2 colonic epithelial cells. *The Korean Journal Parasitology* 52:459-69. doi:10.3347/kjp. 2014.52.5.459.

[70] López-Rosas I, López-Camarillo C, Salinas-Vera YM, Hernández-de la Cruz, Palma-Flores C, Chávez-Munguía B, Resendis-Antonio O, Guillen N, Pérez-Plasencia C, Álvarez-Sánchez ME, Ramírez-Moreno E, Marchat LA. 2019. "*Entamoeba histolytica* up-regulates microRNA-643 to promote apoptosis by targeting XIAP in human epithelial colon cells". *Fronties in Cellular and Infection Microbiology* 8:437. doi.org/10.3389/fcimb.2018.00437.

[71] León-Avila G, Tovar J. (2004). Mitosomes of *Entamoeba histolytica* are abundant mitochondrion-related remnant organelles that lack a detectable organellar genome. *Microbiology* 150:1245-50.doi: 10.1099/mic.0.26923-0.

[72] Mi-ichi F, Makiuchi T, Furukawa A, Sato D, Nozaki T (2011) Sulfate activation in mitosomes plays an important role in the proliferation of *Entamoeba histolytica*. PLoS Negl Trop Dis 5: e1263. doi: 10.1371/journal.pntd.0001263.

[73] Rodriguez MA, Garcia-Perez RM, Mendoza L, Sanchez T, Guillen N, Orozco E (1998) The pyruvate: ferredoxin oxidoreductase enzyme is located in the plasma membrane and in a cytoplasmic structure in *Entamoeba*. Microb Pathog 25: 1–10.

[74] Santos HJ, Hanadate Y, Imai K and Nozaki T. 2019. "An *Entamoeba*-specific mitosomal membrane protein with potential association to the Golgi Apparatus". *Genes.* 10:367. doi:10.3390/genes10050367.

In: *Entamoeba*
Editor: Thomas L. Johnson
ISBN: 978-1-53618-506-5

Chapter 5

STRUCTURAL AND FUNCTIONAL STUDIES OF SERINE BIOSYNTHETIC PATHWAY ENZYMES IN *E. HISTOLYTICA*

Poonam Kumari, Pragyan Parimita Rath and Samudrala Gourinath**[*]**
School of Life Sciences, Jawaharlal Nehru University, New Delhi, India

ABSTRACT

The growth, survival, and virulence of the microaerophilic protozoan *Entamoeba histolytica* are correlated with its redox power and expression of its anti-oxidative enzymes. Thus, the ability to subvert the toxicity of oxidative stress is of central importance for the establishment of successful infection. Cysteine is the major thiol present, and one of the key molecules that help in redox defence of the parasite *E. histolytica*. L-serine is the only substrate for the *de novo* cysteine biosynthesis. D-phosphoglycerate dehydrogenase (PGDH), 3-phosphoserine aminotransferase (PSAT), and phosphoserine phosphatase

[*] Corresponding Author's E-mail: sgourinath@mail.jnu.ac.in.

(PSP) are the three enzymes catalysing the serine synthesis in the so-called phosphorylated pathway from the glycolytic intermediate 3-phosphogly-ceric acid (3-PGA). PGDH catalyses the first committed step of this pathway, and EhPGDH was the first structure to be reported from the Type IIIK PGDH family. The rate-limiting step of serine biosynthesis is carried out by PSP, and *E. histolytica* possesses a novel metal independent PSP, in contrast to its human host. This chapter aims to summarise our current knowledge on the identification, characterisation, structure, and function of the enzymes of serine biosynthesis in *E. histolytica*. The recent findings also reveal that the serine biosynthesis is capable of counteracting the increased cysteine synthesis demand under oxidative stress conditions.

Keywords: cysteine, serine biosynthesis, oxidative stress, *Entamoeba histolytica*

INTRODUCTION

Entamoeba histolytica survives only sparse levels of oxygen, during its average growth and propagation. However, it is an invasive pathogen and gets exposed to high concentrations of oxygen and the other reactive oxygen molecules (ROS) during its attack on cells [1]. Thus, for a successful infection, amoebae must detoxify these harmful compounds by deploying the anti-oxidative machinery for its rescue. These mechanisms play a critical role in the biology of *E. histolytica*.

In this catalase and peroxidase-deficient micro-aerophilic parasite, cysteine is the prime anti-oxidative agent [2, 3]. Cysteine, a principle low molecular weight thiol, is considered to be vital for the growth and survival of *E. histolytica*. It is also a source of the inorganic sulfur atom for the biosynthesis of Fe-S clusters, biotin, Co-A, and various other anti-oxidative thiols (for example, glutathione) [4-6]. Additionally, it is implicated in various pivotal biological processes, including attachment, proliferation, and motility [7, 8].

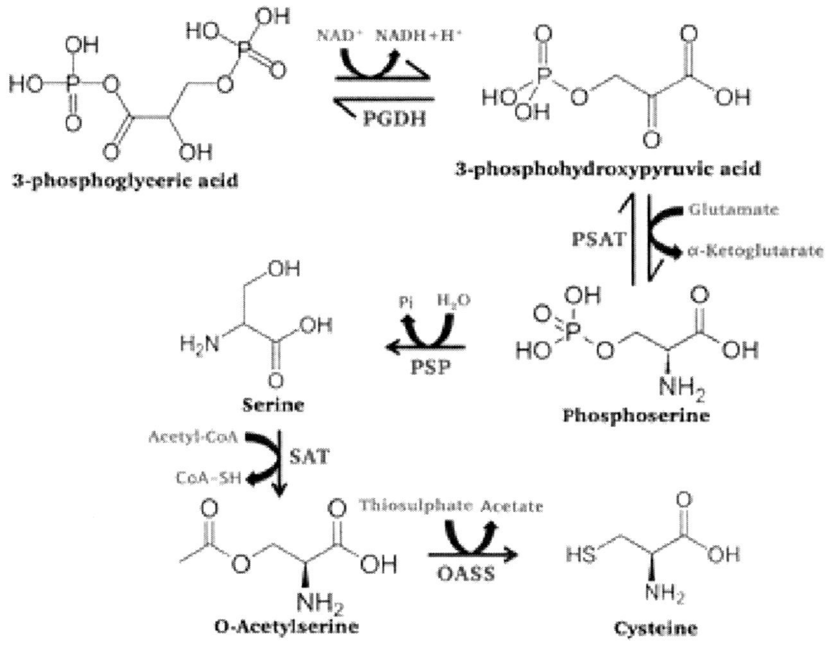

Figure 1. Synthesis of cysteine from serine. Displayed here is the complete pathway beginning from 3-Phosphoglyceric acid en route to the production of cysteine operating in *E. histolytica*. The enzymes catalysing the respective step are highlighted in Blue, while cofactors or other reactant products entering the pathway are in Red. The secondary products released are coloured Green in the diagram.

E. histolytica is cysteine autotrophic and uses the sulfur-assimilatory *de novo* L-cysteine biosynthetic pathway for its production *in vivo*. In this pathogen, two enzymes work in tandem to synthesise cysteine from serine, Serine Acetyl Transferase (SAT) and O-Acetyl Serine Sulfhydrylase (OASS) (Figure 1). Serine is the only substrate for *de novo* cysteine biosynthesis, where it is converted to O-acetyl serine (OAS) by SAT and OAS is further converted to cysteine by OASS. Amoeba carries three isoforms each for these enzymes, and studies have shown that over-expression of OASS gene protects the amoebic trophozoites from oxidative stress [9-14]. The feedback regulation of the SAT has been designed to continually maintain high levels of cysteine in this organism [15]. Cysteine synthesis is regulated by the formation of the cysteine synthase complex in bacteria and plants, unlike *E. histolytica*, where SAT and OASS do not interact with each other [12, 14]. Suppression of these genes is fatal for the survival of this organism, and the addition of inhibitors of SAT or OASS significantly retards the growth of *E. histolytica* [10, 16-18]. Moreover, a decrease in cysteine levels also alters the expression of the genes involved in metabolism, signalling, and DNA/RNA regulation and transport, leading to growth defects in this parasite [19]. Cysteine deprivation, thus, consequently affects this protozoan parasite. However, the *de novo* cysteine synthesis pathway does not exist in humans, opening avenues for targeted drug designing [20].

Besides acting as the prime substrate for cysteine production, serine serves as an elementary unit for the protein synthesis. It is also a key intermediate in a variety of metabolic pathways required for the generation of glycine, L-methionine, purines, porphyrins, phosphatidyl-L-serine, and neuromodulator D-serine [21, 22]. To a great extent, serine is synthesised via two independent routes: the phosphorylated and the non-phosphorylated pathway. In plants and baker's yeast, both these pathways are operative, while, in bacteria and mammals, the phosphorylated pathway predominates [23-25].

In *E. histolytica*, serine is produced from the glycolytic intermediate, 3-phosphoglyceric acid (3-PGA) through the phosphorylated pathway. The pathway is a sequential reaction of three enzymes: Phosphoglycerate dehydrogenase (PGDH, EC 1.1.1.95), Phosphoserine aminotransferase (PSAT, EC 2.6.1.52), and Phosphoserine phosphatase (PSP, EC 3.1.3.3) (Figure 1). PGDH catalyses the NAD^+-dependent oxidation of 3-PGA to phosphohydroxy pyruvic acid (3-PHP) [26]. Next, a transamination reaction, which transfers an amino group from glutamate to 3-PHP, is carried out by the second enzyme of the pathway, PSAT, generating phosphoserine and α-ketoglutarate (AKG) [27]. The last step of the pathway involves the cleavage of phosphoserine by PSP to produce serine. PGDH and PSAT catalyse reversible reactions in contrast to PSP that drives the irreversible de-phosphorylation reaction, serving as the rate-limiting step of the pathway [28, 29]. In general, the pathway is negatively regulated by the feedback inhibition of PGDH by the end product, i.e., serine, suggesting that PGDH plays a central role in the regulation of serine synthesis. However, amoebic PGDH, un-inhibited by serine maintains its high cellular levels, in turn leading to high levels of cysteine in *E. histolytica* [30].

Serine biosynthetic pathway presents itself as the precursor for cysteine metabolism in amoebae (Figure 1), and the pathway exists exclusively in the cytosol. In this parasite, all the three enzymes of serine synthesis are well characterised at molecular, biochemical, structural, and functional levels. Studies have shown that enzymes of this pathway significantly differed from the human host and have a crucial role in the oxidative stress response of *E. histolytica*. Hence, they can be exploited as potential targets for designing drugs for the treatment of amoebiasis.

In this chapter, we chalk out a detailed picture of the structural and functional aspects of all the enzymes operational in the serine biosynthetic pathway in *E. histolytica*. We explain their characteristic behaviour with respect to homologues from other organisms, and also discuss their roles as an anti-oxidant.

D-Phosphoglycerate Dehydrogenase from *Entamoeba histolytica* (EhPGDH)

PGDH catalyses the first committed step of the phosphorylated serine biosynthetic pathway and belongs to the 2-hydroxy acid dehydrogenase family [31]. The domain organisation/complexity describes three types of structural variants from PGDH, i.e., I, II, and III. Type III is the simplest form as it lacks all the regulatory domains, and is further sub-classified into two groups, based on their critical active site residue; either lysine (Type IIIK) or histidine (Type IIIH) [32, 33]. EhPGDH gene encodes a 299 amino acids long Type IIIK enzyme. The amino acid comparisons and phylogenetic analyses demonstrate the high degree of homology between the amoebic PGDH and the PGDH from Bacteroides (Figure 2A) [26].

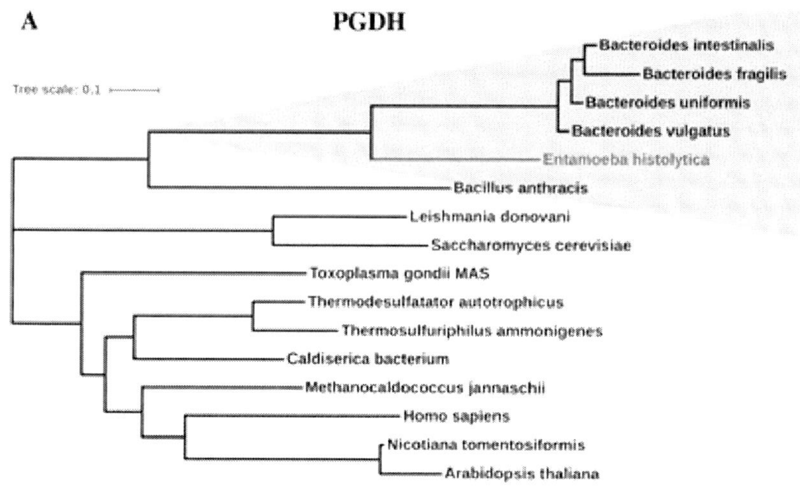

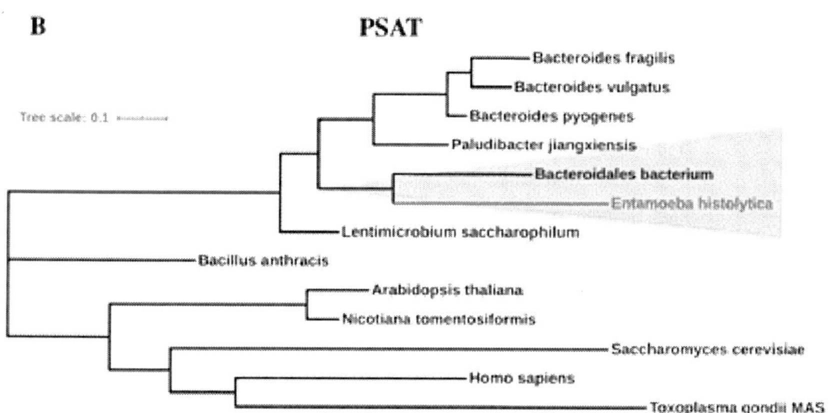

Figure 2. Evolution of PGDH and PSAT enzymes. The phylogenetic tree depicts the evolution course of A. PGDH and B. PSAT enzymes in *E. histolytica*. In both the cases, the amoebic enzymes highlighted in Red branch out with homologues from Bacteroides, showing the higher identity with bacterial proteins.

STRUCTURAL DETAILS OF EHPGDH

Type I and II PGDH are divided into three distinct structural domains: the nucleotide-binding domain (NBD), the substrate-binding domain (SBD), and the regulatory domain [34, 35]. In some organisms like *M.*

tuberculosis, PGDH also has a fourth intervening domain that is present in between the SBD and the regulatory domain [36]. The crystal structure of EhPGDH was the first Type IIIK PGDH structure to be reported. Structural analysis revealed that each protomer of EhPGDH comprises of only two domains: the NBD that binds to the cofactor, and the SBD [37], representing only the functional core of PGDH family, with no additional domain. The N-terminal 95 residues that fold into five parallel β strands and four interconnecting α helices forms the significant part of the SBD. The C-terminal residues (267–299) shape the fifth α-helix. The central β sheet is surrounded by α helices (Figure 3).

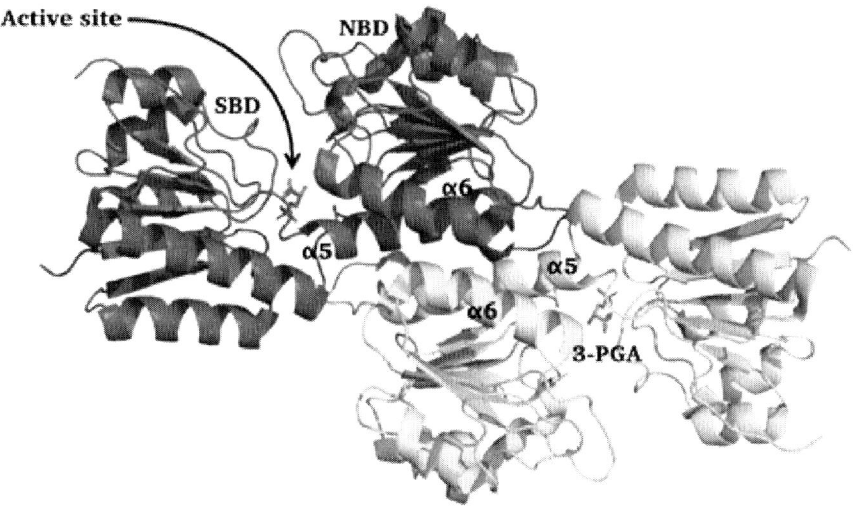

Figure 3. The functional dimer of EhPGDH. The structure comprises of two monomers coloured Red and Yellow respectively, each bound to their substrate, 3-Phosphoglyceric acid (3-PGA). The Substrate-binding domain (SBD) and the Nucleotide-binding domain (NBD) are labelled, enclosing the active site pocket. Both α5 and α6 are marked to show the dimeric interface.

The intervening residues (101–259) of EhPGDH form the NBD containing six central parallel β strands surrounded by six α helices, as shown in Figure 3.

The overall folding pattern of EhPGDH domains is similar to that of the other PGDH structures [34, 36]. The NDB and SBD are loosely connected (involving only two loops) and this thin connection functions as a hinge that permits the domain movement upon substrate binding for effective catalysis. The interface between the two domains forms the active site of EhPGDH, involving residues from both, the cofactor (NAD^+)-binding site on NBD, and the substrate-binding site on SBD (Figure 3). The catalytic triad His/Glu/Arg forms a charge relay system for hydride transfer in the PGDH family [38-40]. This triad is absent in Type IIIK enzymes, such that, a lysine residue replaces histidine, while glutamate is altogether absent from the active site. The crystal structure of EhPGDH in complex with the substrate and the cofactor provided clear evidence for Lys263 as the single residue responsible for an efficient catalytic mechanism, rather than a His/Asp or His/Glu pair, those involved in catalysis of any other PGDH [37]. The biochemical studies also showed that the mutation of Lys263 to Ala led to a complete loss of the enzymatic activity.

EhPGDH Is a Dimeric Enzyme

Type I and Type II PGDH enzymes from bacteria, plants, and mammals stably exist as a tetramer (i.e., dimer of dimers) [36, 41, 42]. In contrast, the amoebic PGDH, a Type III enzyme, exists as a homodimer as demonstrated by gel filtration chromatography and crystallographic studies [26, 37]. EhPGDH lacks (a) the C-terminal serine binding regulatory domain, which is necessary for tetramerisation, and (b) the 13-14 amino acids long region containing the conserved tryptophan residue in the NDB, also essential for tetramerisation [43]. The NDB of two subunits predominantly participates in the dimerisation of EhPGDH, where the α5 and α6 helices of the NBD are the major contributors (Figure 3).

Ligand Binding Induces Domain Movement That Leads to Cleft Closure

The closure of the active site cleft shields it from the solvent and is required for effective catalysis. This also helps in optimal positioning of the reactants close enough to form the product. Insights into the structural changes occurring upon binding of the ligand to the active site of EhPGDH came from 3-PGA-(substrate) and NAD^+-(cofactor) bound structures. The apo-EhPGDH (i.e., unbound state) was captured with an open active site cleft in its crystal structure. Binding of these ligands accompanied a conformational change in the enzyme, where the SBD showed a rotation of about 20° relative to the other domain, thus corresponding to the ligand-induced closure of the catalytic cleft (Figure 4) [37].

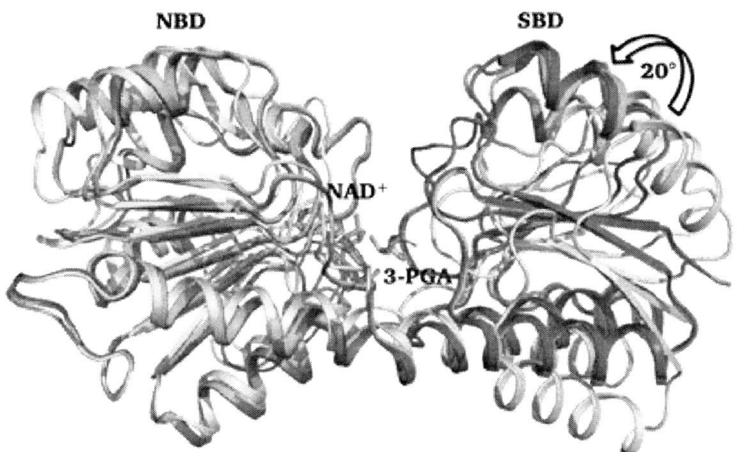

Figure 4. Domain movements initiated during ligand binding. The apo-EhPGDH (Green) was superimposed with both NAD^+ bound (Cyan) and 3-PGA bound (Red) structures. In either case, a 20° movement was observed leading to the closure of the active site. NAD^+ is coloured Blue, while 3-PGA is in Pink to highlight their special occupancy in the structures.

The crystal structures of Type I (from *M. tuberculosis*, MtbPGDH) and Type II PGDH (from *E.coli*, EcPGDH) are also available in complex with the ligands. A comparative study of ligand-bound and unbound MtbPGDH structures yields an r.m.s.d. of 0.16Å, suggesting that there is no significant structural change in the enzyme upon substrate binding and the active site was observed in an open conformation in both the cases [44]. Whereas, the comparison of EcPGDH, apo- and complex structure shows an r.m.s.d. of 2.03Å, indicating that EcPGDH undergoes some conformational changes upon its substrate binding [35].

In contrast to Type I and II enzymes, Type III PGDH shows a significant domain movement for the effective closure of the enzyme's active site during the catalytic cycle. For example, on cofactor binding, *S. tokodaii* PGDH (a Type IIIH enzyme) resembles the EhPGDH complex structure with a closed cleft conformation (PDB ID: 2EKL). As mentioned earlier, Type III is the simplest group, consisting of only the functional core of PGDH, while; Type I and II are complex enzymes with additional regulatory and C-terminal domains. Thus, we conclude from the above analysis that in the PGDH family, the magnitude of domain movement in response to the ligand binding increased as the complexity of the domain organisation decreased.

PHOSPHOSERINE AMINOTRANSFERASE FROM *ENTAMOEBA HISTOLYTICA* (EHPSAT)

PSAT is a vitamin B6-dependent enzyme that belongs to the α-family of pyridoxal-5'-phosphate (PLP) enzymes [45]. From bacteria to plants, the enzyme exists as a homodimer, and this dimeric assembly is essential for its functional activity [46-48]. EhPSAT exists as a polypeptide of 358 amino acid residues. The amino acid sequence alignment and phylogenetic analyses revealed that EhPSAT also has a close kinship with the Bacteroides PSAT, similar to EhPGDH [27]. The close association of EhPGDH and EhPSAT with their bacterial counterparts suggested that *E. histolytica* gained both of these genes from an ancestral organism of Bacteroides by lateral gene transfer (Figure 2B). EhPSAT was characterised by Ali and coworkers, who also confirm its *in vivo* role in the serine biosynthesis by functional rescues of the *E. coli* PSAT-deficient mutant cells [27].

STRUCTURE OF EHPSAT

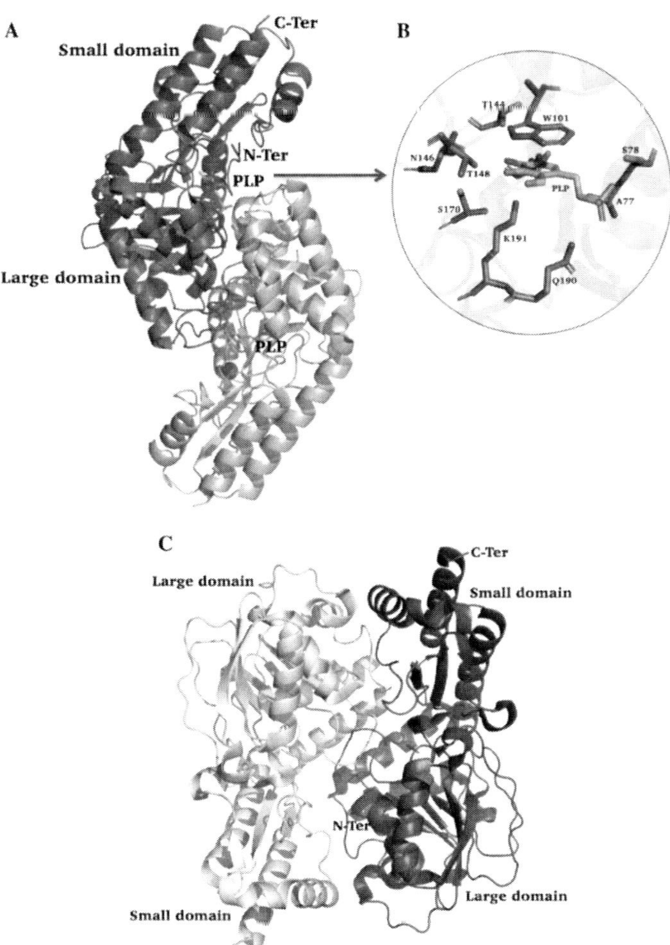

Figure 5. Structural insights of wildtype and mutant EhPSAT. A. The large domain from each monomer of EhPSAT interacts to form the active dimer. Chain A is coloured Deep purple, and Chain B is Golden. B. The zoomed-in view of the active site shows the alignment of the interacting residues with the substrate, PLP. C. The change in the dimerisation pattern is visible in the EhPSATΔ45 mutant.

The overall architecture of EhPSAT is similar to the other family members. Each subunit of EhPSAT predominantly consists of two distinct domains: a large PLP binding domain and another small domain [49]. The residues 15-245 make up the large domain and are organised into the classical seven-stranded β sheet, which is the signature fold of all α-family aminotransferases [46, 47, 50]. This β sheet is surrounded by six α helices on both its sides. The smaller C-terminal domain, a comparatively compact structure is made up of a combination of 100 C-terminal residues and 1-15 N-terminal residues, containing a three-stranded anti-parallel β-sheet surrounded by three α helices (Figure 5A) [49].

THE FUNCTIONAL UNIT OF EHPSAT IS A DIMER

EhPSAT crystallised as a homodimer consistent with the solution state. Interactions between the two monomers are quite extensive, involving residues from both the large and the small domains of each monomer. However, the large domain contributes to the majority of the residues (Figure 5A). Structural and biochemical studies have shown that the N-terminal residues are crucial for its quaternary structure, the active conformation of EhPSAT. The EhPSATΔ45 mutant (45 residues deleted from its N-terminal) also crystallised in a dimeric state like the wild type (WT) protein; however, the arrangement of the two monomers was very different in this case (Figure 5C). To make a dimer, the second protomer of the mutant rotates approximately 80Å, compared to the protomer of WT protein, suggesting the importance of N-terminal residues in the active conformation of EhPSAT [49].

PLP is required as an essential cofactor for all the known PSAT proteins to date [51, 52]. One PLP molecule per monomer of EhPSAT was also observed in the crystal structure. However, the active site of the EhPSATΔ45 mutant structure lacked any PLP. Moreover, the mutant also showed a complete loss of enzymatic activity, suggesting that the deletion of 45 N-terminal residues led to the formation of a non-functional dimer. This observation is consistent with the earlier study reported by Mishra et al., who described the importance of the N- and the C-terminal residues in the stabilisation of dimeric conformation of EhPSAT [53].

THE ACTIVE SITE

The EhPSAT dimer has two structurally identical active sites situated on either side of the dimeric interface (Figure 5A). In each of the active site, a PLP molecule is bound through an aldimine linkage with Lys191. Crystallographic studies of *E. coli* PSAT have shown a close association of PLP molecule with the Trp, a strictly conserved residue found in all PSATs [46].Similarly, in the EhPSAT structure, a prominent stacking interaction is observed between the pyridoxal ring of the cofactor and the indole ring of Trp101 (Figure 5B). Moreover, these interactions and the orientation of the bound PLP molecule in the active site of EhPSAT match the other reported PSAT structures [47, 48, 54, 55]. Mutation of Trp101 to either His, Phe, or Ala resulted in significant loss of the enzymatic activity [56]. This study justifies the conservation of Trp in both the functional activity and the stability of EhPSAT.

EHPGDH AND EHPSAT DISPLAY NOVEL INTERACTION MODE

The metabolic pathways are composed of hundreds of interconnected chemical reactions. For the regulation of these complex metabolic processes, cells have developed few strategies; for example, enzymes show higher efficiency when they work in an organised complex. Such type of interactions also led to channelling of the substrate and the intermediates from one catalytic site to another, without diffusing into the bulk phase, due to the proximity of the active site of the interacting partners. In *E. histolytica*, the first two enzymes (i.e., EhPGDH and EhPSAT) of phosphorylated serine biosynthetic pathway are observed to interact. They interact specifically through the nucleotide-binding domain of EhPGDH and PLP binding domain of EhPSAT. In case of EhPGDH-EhPSAT complex, the value of Km for the substrate 3-PGA was found significantly lower when compared to the individual PGDH enzyme, indicating that the product of PDGH reaction is being taken up by its interacting partner [57].

PHOSPHOSERINE PHOSPHATASE FROM *ENTAMOEBA HISTOLYTICA* (EHPSP)

As mentioned above, the rate-limiting step of the phosphorylated pathway is the hydrolysis of L-phosphoserine by Phosphoserine phosphatase (PSP). PSPs are found and studied from all the three domains of life: Bacteria, Archaea, and Eukarya. All known PSPs are Mg^{2+} dependent enzymes and belong to the haloacid dehalogenase-like hydrolase superfamily [58-60]. In *E. histolytica*, the first two enzymes of this pathway, PGDH and PSAT were characterised long back, but PSP was unknown until recently. Therefore, the absence of the classical PSP orthologue in this parasite presented a missing link in serine anabolism.

Recently, PSP was identified and characterised in *E. histolytica* (EhPSP) from our lab that belongs to the histidine phosphatase (HP) superfamily [29], which was earlier annotated as a probable cofactor dependent Phosphoglycerate mutase (dPGM) protein (AmoebaDB ID: EHI_129820). Further, using various molecular, structural, and functional approaches, it was established as an actual PSP instead of a dPGM. The superfamily mentioned above consists of several types of proteins, including dPGMs and phosphatases, since both the types of proteins share several catalytic residues. Contrary to the other classical PSPs (including human PSP), EhPSP is a metal-independent enzyme. *E. histolytica* has a single isoform of PSP made up of 205 amino acids [29].

EhPSP: Structural Insights

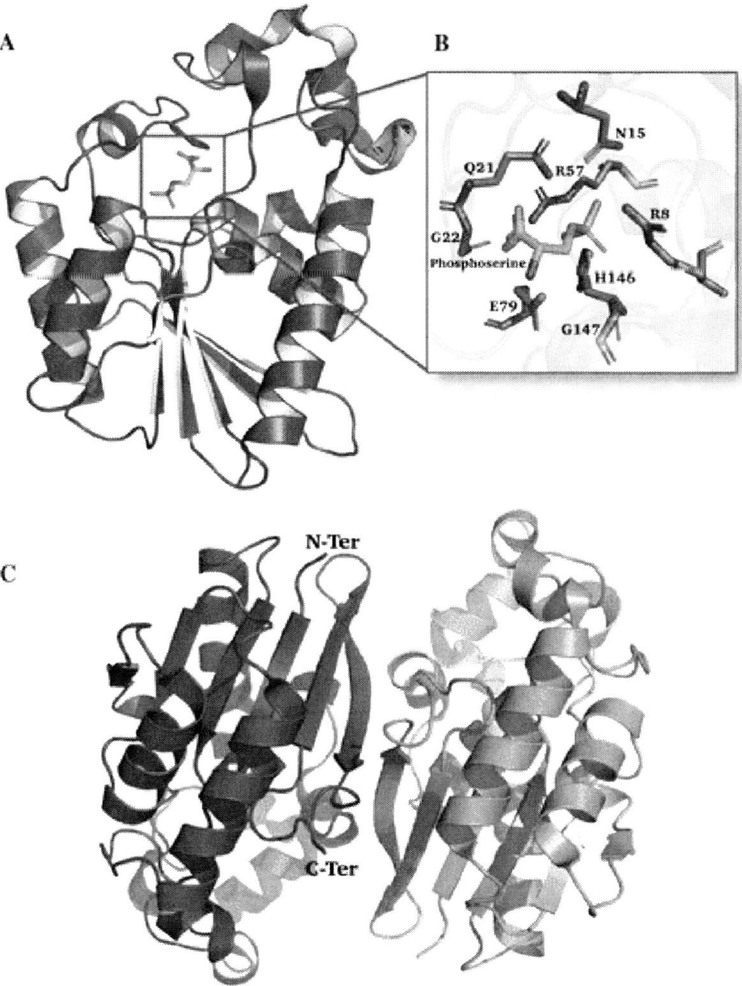

Figure 6. Structure of EhPSP. A. Phosphoserine bound EhPSP. B. The residues of the active site are labelled in interaction with the substrate, phosphoserine. C. The functional dimer of EhPSP formed through the interaction of β sheets in the middle.

EhPSP is a single domain enzyme with the α/β/α fold, consisting of eight α helices, and a central five-stranded mixed β sheet. Four α helices surround the sheet on either side with an additional small sub-domain containing two α-helices (Figure 6A). This arrangement is the signature architecture of the histidine phosphatase superfamily proteins. Additionally, a phosphate ion was bound in the active site of EhPSP, as seen in its crystal structure. The phosphate ion can occupy either the substrate (phosphoserine) or the product phosphate positions. The functional unit of EhPSP is a dimer, and at the dimeric interface, a continuous ten-stranded anti-parallel β sheet can be seen. The terminal β strand and several C-terminal residues from each of the monomer participate in the dimerisation (Figure 6C) [29]. In summary, the overall architecture and dimerisation pattern of EhPSP was quite similar to the other phosphatases in the superfamily [61-63]. EhPSP is only 20% identical to the human homologue since they both belong to two different families. When compared, an r.m.s. deviation of 5.1 Å for 72 Cα atoms was observed for these structures, revealing the high degree of structural differences [28].

THE CONSERVED ACTIVE SITE POCKET

Catalytic activities of the enzymes in the HP superfamily rely on a conserved histidine residue present in the characteristic "RHG motif" at the N-terminus of the protein. This histidine mediates the reaction by undergoing phosphorylation and de-phosphorylation during catalysis [64]. A conserved phosphate pocket marks the residues that interact with the phosphate moiety of their substrate during catalysis including a pair of arginine residues and another histidine residue at the C-terminal of the protein (Arg7, Arg55, and His108 in *E.coli* SixA protein) [65]. The central β sheet on one end flanks the active site of EhPSP while the other end is closed by two loops. Residues from different parts of the protein comes together to form the active site (Figure 6B). Biochemical studies have shown that an active site mutant of EhPSP, where His9 from RHG motif is replaced with Ala, displayed only 4% of the WT activity. Thus, like the other family members, His9 serves as a base during the EhPSP activity.

Furthermore, the crystal structure of the mutant EhPSP in complex with its substrate phosphoserine helped in deciphering the interacting catalytic residues (Figure 6B). A comparative analysis of the active site of EhPSP with the phosphatases from the superfamily revealed a highly conserved phosphate pocket [29].

PHOSPHOSERINE PHOSPHATASE SAFEGUARDS *ENTAMOEBA HISTOLYTICA* DURING OXIDATIVE STRESS

The oxidative stress (OS) response plays guard during immune evasion by *E. histolytica*, leading to successful infection. Cysteine protects the parasite from the host-derived reactive oxygen species throughout the OS response. Previously, cysteine synthase (CS) over-expressing amoebic trophozoites were observed to be more resistant to exogenous hydrogen peroxide (H_2O_2) [14]. CS catalyses the production of S-methyl cysteine (SMC), and the recent studies have shown that SMC protects *E. histolytica* from H_2O_2 induced oxidative stress. When the CS gene was down-regulated, the trophozoites became more sensitive to stress [66].

We have recently enlightened the importance of EhPSP, the regulatory enzyme in the serine biosynthesis in the OS response of amoebic trophozoites. Several results have been enlisted as evidence for the same. Firstly, when the WT EhPSP was over-expressed, the amoebic trophozoites developed resistance to H_2O_2-triggered stress. However, such resistance was not observed in cells expressing the functional mutant (EhPSP-His9 to Ala) of the protein. Secondly, the over-expression of EhPSP (WT) also led to the reduction of ROS level in the amoebic cells. This effect was reversed in trophozoites over-expressing the functional mutant of EhPSP. There was a significant increase in the levels of intracellular ROS, correlating well with the effect of EhPSP on the viability of *E. histolytica* under oxidative stress. Besides, under oxidative stress conditions, an increase in the level of EhPSP protein was also observed, to meet the increased demand for cysteine synthesis [29]. In conclusion, these analyses demonstrated that overexpression of enzymes of cysteine biosynthetic pathway (i.e., CS or PSP) confers a survival advantage to this protozoan parasite from oxidative stress (Figure 7).

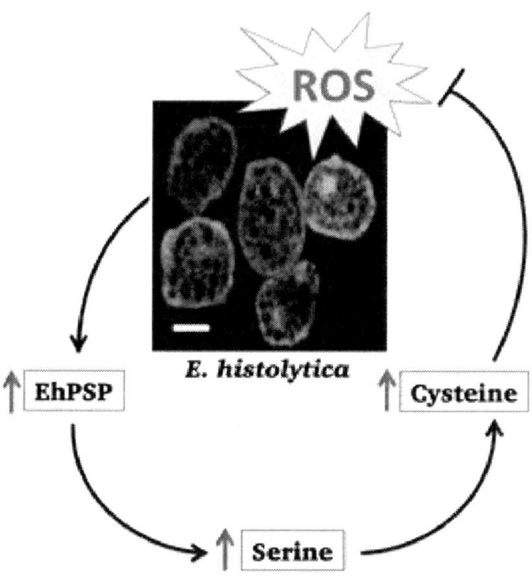

Figure 7. Regulation of cysteine production during oxidative stress. Amoebic trophozoites are shown here captured through confocal microscopy. The cytosolic localisation of EhPSP is shown in Green, while Blue represents the nucleus. When *E. histolytica* is attacked by Reactive oxygen species (ROS), it overexpresses EhPSP, leading to higher production of serine. This, in turn, boosts the cysteine formation that protects the amoebic cells from ROS.

REGULATION OF SERINE BIOSYNTHETIC PATHWAY IN *ENTAMOEBA HISTOLYTICA*

E. histolytica profoundly depends on the glycolytic pathway for its energy requirements, since it lacks the other major energy-generating pathways, such as oxidative phosphorylation and TCA cycles. As mentioned earlier, the phosphoserine biosynthetic pathway utilises 3-PGA (a glycolytic intermediate) as a substrate. Therefore, to maintain cellular energy homeostasis, this pathway must be tightly regulated. On the other hand, to meet the higher demand of cysteine, the parasite must synthesise serine accordingly. With this purpose, a few strategies are designed to function at various levels in this pathogen.

Firstly, to regulate the serine biosynthesis in bacteria and plants, the first enzyme of this pathway, that is, PGDH is allosterically inhibited by the end product of the pathway, serine [67-69]. However, contrasting to its counterpart in *M. tuberculosis* and *E.coli* [36, 70, 71], the amoebic PGDH is insensitive to serine induced inhibition because it lacks the serine-binding regulatory domain [72]. This indicates that this protozoan parasite has adopted a non-feedback regulatory enzyme, in order to produce enough serine to fuel cysteine production to protect itself from oxidative stress during infection.

Secondly, the structure analysis of EhPSAT revealed one chloride ion bound to the active site. Previously, iodide and chloride ions were also present in PSAT from *Trichomonas vaginalis* and *Bacillus alcalophilus* [48, 50]. Biochemical studies have shown that few halide ions, such as NaCl and NaBr, inhibit the enzymatic activity of EhPSAT in the forward direction, i.e., PHP to phosphoserine formation [49, 73]. However, the activity of EhPSAT in the reverse direction, i.e., synthesis of PHP from phosphoserine is not inhibited by any halides. These analyses strongly conclude that *Entamoeba* exercises tight regulation of OPLS synthesis and thus the formation of serine but not of PHP, necessary for coordination of the consumption of glycolytic intermediates.

CONCLUSION

Microaerophilic pathogens encounter oxidative stress as a significant roadblock while infecting their hosts. *E. histolytica* utilises cysteine as the prime anti-oxidant molecule, unlike the classical molecules usually recruited by other aerobic organisms. Such adaptive response provides a shield to this protozoan parasite against the host immune defence attack, contributing significantly to its pathogenic potential. Serine is essential substrate for cysteine synthesis. This chapter enlightens the detailed structural and functional aspects of the phosphorylated serine biosynthetic pathway enzymes working in *E. histolytica*.

EhPGDH exists as a dimer both in solution and in crystals, contrary to the PGDH from bacteria and plants. The structure of EhPGDH represents the functional core of the PGDH family. EhPGDH lacks the C-terminal regulatory domain and thus does not display the feedback inhibition by serine. Hence, *Entamoeba* has acquired a strategy to bypass the regulatory mechanisms of serine synthesis for the higher production of serine/ cysteine. *E. coli* cells can be complemented with EhPSAT to recover their function. Both EhPGDH and EhPSAT are evolutionarily close to their bacterial counterparts, suggesting that *E. histolytica* has acquired both of these genes by lateral gene transfer from Bacteroides. These amoebic proteins interact with each other as demonstrated by their biochemical and *in vitro* interaction studies to efficiently synthesise phosphoserine.

A recent study reported the existence of functional PSP in the amoebic genome. Unlike the classical PSPs, EhPSP is a metal independent enzyme that belongs to a novel PSP enzyme category. It is found to be completely different from its *M. tuberculosis* counterpart. Moreover, out of the three amitochondriate parasites, i.e., *Entamoeba histolytica*, *Trichomonas vaginalis*, and *Giardia lamblia*, PSP has been identified only in *E. histolytica* so far. Lastly, the literature garnered from the *in vivo* studies indicates an increase in the level of EhPSP when trophozoites are subjected to oxidative stress. They counteract the increased cysteine synthesis demand by over-expressing EhPSP under such conditions. The pivotal role of EhPSP in the anti-oxidative defence mechanism of *E. histolytica* and its absence in the human host identifies this protein as a novel drug target aimed at combating amoebiasis.

Once serine is formed, SAT and OASS catalyse the formation of cysteine.The studies conducted on the cysteine biosynthetic pathway unravelled differential strategies appointed by *E. histolytica* to survive in their human host. Since all the three isoforms had varying inhibitory controls by feedback inhibition, this organism ensures a constant production of cysteine *in vivo*. Generally, this pathway is also regulated by the association of SAT and OASS enzymes to form the cysteine synthase (CS) complex. While in *E. histolytica* these proteins do not interact and do not form CS complex to maintain a higher concentration of the crucial cystein under all conditions. Amoebic SAT C-terminal and the active site of OASS were modified, resulting in loss of the CS complex formation, a regulatory mechanism used by other organisms to escape the complete inhibition of the cysteine biosynthesis. Thus, the complete pathway beginning from 3-phosphoglycerate till the final production of cysteine has been tailor-made to protect *E. histolytica* from the oxidative stress encountered during the infection.

REFERENCES

[1] Stanley Jr SL. Amoebiasis. *The lancet.* 2003;361(9362):1025-34.

[2] Krauth-Siegel RL, Leroux AE. Low-molecular-mass antioxidants in parasites. *Antioxidants & redox signaling.* 2012;17(4):583-607.

[3] Fahey RC, Newton GL, Arrick B, Overdank-Bogart T, Aley SB. *Entamoeba histolytica*: a eukaryote without glutathione metabolism. *Science.* 1984;224(4644):70-2.

[4] Ali V, Nozaki T. Current therapeutics, their problems, and sulfur-containing-amino-acid metabolism as a novel target against infections by "amitochondriate" protozoan parasites. *Clinical microbiology reviews.* 2007;20(1):164-87.

[5] Ali V, Shigeta Y, Tokumoto U, Takahashi Y, Nozaki T. An intestinal parasitic protist, *Entamoeba histolytica*, possesses a non-redundant nitrogen fixation-like system for iron-sulfur cluster assembly under anaerobic conditions. *Journal of Biological Chemistry.* 2004;279(16):16863-74.

[6] Kessler D. Enzymatic activation of sulfur for incorporation into biomolecules in prokaryotes. *FEMS microbiology reviews.* 2006;30(6):825-40.

[7] Gillin FD, Diamond LS. *Entamoeba histolytica* and Giardia lamblia: effects of cysteine and oxygen tension on trophozoite attachment to

glass and survival in culture media. *Experimental parasitology.* 1981;52(1):9-17.

[8] GILLIN FD, DIAMOND LS. Attachment of *Entamoeba histolytica* to glass in a defined maintenance medium: specific requirement for cysteine and ascorbic acid. *The Journal of protozoology.* 1980;27(4):474-8.

[9] Chinthalapudi K, Kumar M, Kumar S, Jain S, Alam N, Gourinath S. Crystal structure of native O-acetyl-serine sulfhydrylase from *Entamoeba histolytica* and its complex with cysteine: structural evidence for cysteine binding and lack of interactions with serine acetyl transferase. *Proteins.* 2008;72(4):1222-32. doi: 10.1002/prot. 22013. PubMed PMID: 18350570.

[10] Dharavath S, Vijayan R, Kumari K, Tomar P, Gourinath S. Crystal structure of O-Acetylserine sulfhydralase (OASS) isoform 3 from *Entamoeba histolytica*: Pharmacophore-based virtual screening and validation of novel inhibitors. *Eur J Med Chem.* 2020;192:112157. doi: 10.1016/j.ejmech.2020.112157. PubMed PMID: 32145643.

[11] Kumar S, Mazumder M, Dharavath S, Gourinath S. Single residue mutation in active site of serine acetyltransferase isoform 3 from *Entamoeba histolytica* assists in partial regaining of feedback inhibition by cysteine. *PLoS One.* 2013;8(2):e55932. doi: 10.1371/journal.pone.0055932. PubMed PMID: 23437075; PubMed Central PMCID: PMCPMC3578862.

[12] Kumar S, Raj I, Nagpal I, Subbarao N, Gourinath S. Structural and biochemical studies of serine acetyltransferase reveal why the parasite *Entamoeba histolytica* cannot form a cysteine synthase complex. *J Biol Chem.* 2011;286(14):12533-41. doi: 10.1074/jbc. M110.197376. PubMed PMID: 21297164; PubMed Central PMCID: PMCPMC3069455.

[13] Raj I, Mazumder M, Gourinath S. Molecular basis of ligand recognition by OASS from *E. histolytica*: insights from structural and molecular dynamics simulation studies. *Biochim Biophys Acta.* 2013;1830(10):4573-83. doi: 10.1016/j.bbagen.2013.05.041. PubMed PMID: 23747298.

[14] Nozaki T, Asai T, Sanchez LB, Kobayashi S, Nakazawa M, Takeuchi T. Characterization of the gene encoding serine acetyltransferase, a regulated enzyme of cysteine biosynthesis from the protist parasites *Entamoeba histolytica* and *Entamoeba dispar*. Regulation and possible function of the cysteine biosynthetic pathway in *Entamoeba*. *J Biol Chem.* 1999;274(45):32445-52. doi: 10.1074/jbc.274.45.32445. PubMed PMID: 10542289.

[15] Hussain S, Ali V, Jeelani G, Nozaki T. Isoform-dependent feedback regulation of serine O-acetyltransferase isoenzymes involved in L-cysteine biosynthesis of *Entamoeba histolytica*. *Mol Biochem Parasitol.* 2009;163(1):39-47. doi: 10.1016/j.molbiopara.2008.09.006. PubMed PMID: 18851994.

[16] Agarwal SM, Jain R, Bhattacharya A, Azam A. Inhibitors of Escherichia coli serine acetyltransferase block proliferation of *Entamoeba histolytica* trophozoites. *Int J Parasitol.* 2008;38(2):137-41. doi: 10.1016/j.ijpara.2007.09.009. PubMed PMID: 17991467.

[17] Jeelani G, Sato D, Soga T, Nozaki T. Genetic, metabolomic and transcriptomic analyses of the de novo L-cysteine biosynthetic pathway in the enteric protozoan parasite *Entamoeba histolytica*. *Sci Rep.* 2017;7(1):15649. doi: 10.1038/s41598-017-15923-3. PubMed PMID: 29142277; PubMed Central PMCID: PMCPMC5688106.

[18] Nagpal I, Raj I, Subbarao N, Gourinath S. Virtual screening, identification and in vitro testing of novel inhibitors of O-acetyl-L-serine sulfhydrylase of *Entamoeba histolytica*. *PLoS One.* 2012;7 (2):e30305. doi: 10.1371/journal.pone.0030305. PubMed PMID: 22355310; PubMed Central PMCID: PMCPMC3280239.

[19] Husain A, Jeelani G, Sato D, Nozaki T. Global analysis of gene expression in response to L-Cysteine deprivation in the anaerobic protozoan parasite *Entamoeba histolytica*. *BMC genomics.* 2011;12(1):275.

[20] Mori M, Tsuge S, Fukasawa W, Jeelani G, Nakada-Tsukui K, Nonaka K, et al. Discovery of Antiamebic Compounds That Inhibit Cysteine Synthase From the Enteric Parasitic Protist *Entamoeba histolytica* by Screening of Microbial Secondary Metabolites. *Front*

Cell Infect Microbiol. 2018;8:409. doi: 10.3389/fcimb.2018.00409. PubMed PMID: 30568921; PubMed Central PMCID: PMCPMC 6290340.

[21] Snyder SH, Kim PM. D-amino acids as putative neurotransmitters: focus on D-serine. *Neurochemical research.* 2000;25(5):553-60.

[22] Snell K. Enzymes of serine metabolism in normal, developing and neoplastic rat tissues. *Advances in enzyme regulation.* 1984;22:325-400.

[23] Melcher K, Rose M, Künzler M, Braus GH, Entian K-D. Molecular analysis of the yeast SER1 gene encoding 3-phosphoserine aminotransferase: regulation by general control and serine repression. *Current genetics.* 1995;27(6):501-8.

[24] Ros R, Muñoz-Bertomeu J, Krueger S. Serine in plants: biosynthesis, metabolism, and functions. *Trends in plant science.* 2014;19(9): 564-9.

[25] Walsh D, Sallach H. Comparative studies on the pathways for serine biosynthesis in animal tissues. *Journal of Biological Chemistry.* 1966;241(17):4068-76.

[26] Ali V, Hashimoto T, Shigeta Y, Nozaki T. Molecular and biochemical characterization of d-phosphoglycerate dehydrogenase from *Entamoeba histolytica*: A unique enteric protozoan parasite that possesses both phosphorylated and nonphosphorylated serine metabolic pathways. *European journal of biochemistry.* 2004; 271(13):2670-81.

[27] Ali V, Nozaki T. Biochemical and functional characterization of phosphoserine aminotransferase from *Entamoeba histolytica*, which possesses both phosphorylated and non-phosphorylated serine metabolic pathways. *Molecular and biochemical parasitology.* 2006;145(1):71-83.

[28] Peeraer Y, Rabijns A, Verboven C, Collet J-F, Van Schaftingen E, De Ranter C. High-resolution structure of human phosphoserine phosphatase in open conformation. *Acta Crystallographica Section D: Biological Crystallography.* 2003;59(6):971-7.

[29] Kumari P, Babuta M, Bhattacharya A, Gourinath S. Structural and functional characterisation of phosphoserine phosphatase, that plays critical role in the oxidative stress response in the parasite *Entamoeba histolytica*. *J Struct Biol.* 2019;206(2):254-66. doi: 10.1016/j.jsb.2019.03.012. PubMed PMID: 30935984.

[30] Ali V, Hashimoto T, Shigeta Y, Nozaki T. Molecular and biochemical characterization of D-phosphoglycerate dehydrogenase from *Entamoeba histolytica*. A unique enteric protozoan parasite that possesses both phosphorylated and nonphosphorylated serine metabolic pathways. *Eur J Biochem.* 2004;271(13):2670-81. doi: 10.1111/j.1432-1033.2004.04195.x. PubMed PMID: 15206932.

[31] Grant GA. A new family of 2-hydroxyacid dehydrogenases. *Biochemical and biophysical research communications.* 1989;165 (3):1371-4.

[32] Grant GA. Contrasting catalytic and allosteric mechanisms for phosphoglycerate dehydrogenases. *Archives of biochemistry and biophysics.* 2012;519(2):175-85.

[33] Grant GA. D-3-Phosphoglycerate Dehydrogenase. *Front Mol Biosci.* 2018;5:110. doi: 10.3389/fmolb.2018.00110. PubMed PMID: 30619878; PubMed Central PMCID: PMCPMC6300728.

[34] Schuller DJ, Grant GA, Banaszak LJ. The allosteric ligand site in the V max-type cooperative enzyme phosphoglycerate dehydrogenase. *Nature structural biology*. 1995;2(1):69-76.

[35] Thompson JR, Bell JK, Bratt J, Grant GA, Banaszak LJ. V max regulation through domain and subunit changes. The active form of phosphoglycerate dehydrogenase. *Biochemistry.* 2005;44(15):5763-73.

[36] Dey S, Grant GA, Sacchettini JC. Crystal Structure of Mycobacterium tuberculosis D-3-Phosphoglycerate Dehydrogenase Extreme Asymmetry in a Tetramer of Identical Subunits. *Journal of Biological Chemistry.* 2005;280(15):14892-9.

[37] Singh RK, Raj I, Pujari R, Gourinath S. Crystal structures and kinetics of Type III 3-phosphoglycerate dehydrogenase reveal catalysis by lysine. *The FEBS journal.* 2014;281(24):5498-512.

[38] Lamzin VS, Dauter Z, Wilson KS. Dehydrogenation through the looking-glass. *Nat Struct Biol.* 1994;1(5):281-2. doi: 10.1038/nsb0594-281. PubMed PMID: 7664032.

[39] Birktoft JJ, Banaszak LJ. The presence of a histidine-aspartic acid pair in the active site of 2-hydroxyacid dehydrogenases. X-ray refinement of cytoplasmic malate dehydrogenase. *J Biol Chem.* 1983;258(1):472-82. doi: 10.2210/pdb2mdh/pdb. PubMed PMID: 6848515.

[40] Taguchi H, Ohta T. Essential role of arginine 235 in the substrate-binding of Lactobacillus plantarum D-lactate dehydrogenase. *J Biochem.* 1994;115(5):930-6. doi: 10.1093/oxfordjournals.jbchem.a124441. PubMed PMID: 7961609.

[41] Grant GA, Kim SJ, Xu XL, Hu Z. The Contribution of Adjacent Subunits to the Active Sites ofD-3-Phosphoglycerate Dehydrogenase. *Journal of Biological Chemistry.* 1999;274(9):5357-61.

[42] Achouri Y, RIDER MH, Van Schaftingen E, Robbi M. Cloning, sequencing and expression of rat liver 3-phosphoglycerate dehydrogenase. *Biochemical Journal.* 1997;323(2):365-70.

[43] Grant GA, Xu XL. Probing the Regulatory Domain Interface ofd-3-Phosphoglycerate Dehydrogenase with Engineered Tryptophan Residues. *Journal of Biological Chemistry.* 1998;273(35):22389-94.

[44] Dey S, Burton RL, Grant GA, Sacchettini JC. Structural analysis of substrate and effector binding in Mycobacterium tuberculosis D-3-phosphoglycerate dehydrogenase. *Biochemistry.* 2008;47(32):8271-82.

[45] Alexander FW, Sandmeier E, Mehta PK, Christen P. Evolutionary relationships among pyridoxal-5′-phosphate-dependent enzymes: Regio-specific α, β and γ families. *European journal of biochemistry.* 1994;219(3):953-60.

[46] Hester G, Stark W, Moser M, Kallen J, Marković-Housley Z, Jansonius JN. Crystal structure of phosphoserine aminotransferase from Escherichia coli at 2.3 Å resolution: comparison of the unligated enzyme and a complex with α-methyl-L-glutamate. *Journal of molecular biology.* 1999;286(3):829-50.

[47] Sekula B, Ruszkowski M, Dauter Z. Structural analysis of phosphoserine aminotransferase (Isoform 1) from Arabidopsis thaliana–the enzyme involved in the phosphorylated pathway of serine biosynthesis. *Frontiers in plant science.* 2018;9:876.

[48] Singh RK, Mazumder M, Sharma B, Gourinath S. Structural investigation and inhibitory response of halide on phosphoserine aminotransferase from Trichomonas vaginalis. *Biochimica et Biophysica Acta (BBA)-General Subjects.* 2016;1860(7):1508-18.

[49] Singh RK, Tomar P, Dharavath S, Kumar S, Gourinath S. N-terminal residues are crucial for quaternary structure and active site conformation for the phosphoserine aminotransferase from enteric human parasite *E. histolytica. International journal of biological macromolecules.* 2019;132:1012-23.

[50] Dubnovitsky AP, Kapetaniou EG, Papageorgiou AC. Enzyme adaptation to alkaline pH: atomic resolution (1.08 Å) structure of phosphoserine aminotransferase from *Bacillus alcalophilus. Protein Science.* 2005;14(1):97-110.

[51] Basurko MJ, Marche M, Darriet M, Cassaigne A. Phosphoserine aminotransferase, the second step-catalyzing enzyme for serine biosynthesis. *IUBMB life.* 1999;48(5):525-9.

[52] John RA. Pyridoxal phosphate-dependent enzymes. *Biochimica et Biophysica Acta (BBA)-Protein Structure and Molecular Enzymology.* 1995;1248(2):81-96.

[53] Mishra V, Ali V, Nozaki T, Bhakuni V. Biophysical characterization of *Entamoeba histolytica* phosphoserine aminotransferase (EhPSAT): role of cofactor and domains in stability and subunit assembly. *European Biophysics Journal.* 2011;40(5):599-610.

[54] Coulibaly F, Lassalle E, Baker HM, Baker EN. Structure of phosphoserine aminotransferase from Mycobacterium tuberculosis. *Acta Crystallographica Section D: Biological Crystallography.* 2012;68(5):553-63.

[55] Kallen J, Kania M, Markovic-Housley Z, Vincent M, Jansonius J. Crystallographic and Solution Studies on Phosphoserine

Aminotransferase (PSAT) from *E. coli. Biochemistry of vitamin B6:* Springer; 1987. p. 157-60.

[56] Mishra V, Kumar A, Ali V, Nozaki T, Zhang KY, Bhakuni V. Role of conserved active site tryptophan-101 in functional activity and stability of phosphoserine aminotransferase from an enteric human parasite. *Amino acids.* 2012;43(1):483-91.

[57] Mishra V, Kumar A, Ali V, Nozaki T, Zhang KY, Bhakuni V. Novel protein–protein interactions between *Entamoeba histolytica* d-phosphoglycerate dehydrogenase and phosphoserine aminotransferase. *Biochimie.* 2012;94(8):1676-86.

[58] Shetty KT. Phosphoserine phosphatase of human brain: partial purification, characterization, regional distribution, and effect of certain modulators including psychoactive drugs. *Neurochemical research.* 1990;15(12):1203-10.

[59] Grant GA. Regulatory Mechanism of Mycobacterium tuberculosis Phosphoserine Phosphatase SerB2. *Biochemistry.* 2017;56(49):6481-90.

[60] Ho C-L, Saito K. Molecular biology of the plastidic phosphorylated serine biosynthetic pathway in *Arabidopsis thaliana. Amino acids.* 2001;20(3):243-59.

[61] Chiba Y, Horita S, Ohtsuka J, Arai H, Nagata K, Igarashi Y, et al. Structural units important for activity of a novel-type phosphoserine phosphatase from *Hydrogenobacter thermophilus* TK-6 revealed by crystal structure analysis. *Journal of Biological Chemistry.* 2013;288(16):11448-58.

[62] Rigden DJ, Mello LV, Setlow P, Jedrzejas MJ. Structure and mechanism of action of a cofactor-dependent phosphoglycerate mutase homolog from *Bacillus stearothermophilus* with broad specificity phosphatase activity1. *Journal of molecular biology.* 2002;315(5):1129-43.

[63] Zheng Q, Jiang D, Zhang W, Zhang Q, Zhao Q, Jin J, et al. Mechanism of dephosphorylation of glucosyl-3-phosphoglycerate by a histidine phosphatase. *Journal of Biological Chemistry.* 2014;289(31):21242-51.

[64] Rigden DJ. The histidine phosphatase superfamily: structure and function. *Biochemical Journal.* 2008;409(2):333-48.

[65] Hamada K, Kato M, Shimizu T, Ihara K, Mizuno T, Hakoshima T. Crystal structure of the protein histidine phosphatase SixA in the multistep His-Asp phosphorelay. *Genes to Cells.* 2005;10(1):1-11.

[66] Husain A, Sato D, Jeelani G, Mi-ichi F, Ali V, Suematsu M, et al. Metabolome analysis revealed increase in S-methylcysteine and phosphatidylisopropanolamine synthesis upon L-cysteine deprivation in the anaerobic protozoan parasite *Entamoeba histolytica. Journal of Biological Chemistry.* 2010;285(50):39160-70.

[67] Sugimoto E, Pizer LI. The mechanism of end product inhibition of serine biosynthesis I. Purification and kinetics of phosphoglycerate dehydrogenase. *Journal of Biological Chemistry.* 1968;243(9):2081-9.

[68] SASKI R, PIZER LI. Regulatory properties of purified 3-phosphoglycerate dehydrogenase from *Bacillus subtilis. European journal of biochemistry.* 1975;51(2):415-27.

[69] Slaughter J, Davies D. Inhibition of 3-phosphoglycerate dehydrogenase by l-serine. *Biochemical Journal.* 1968;109(5):749-55.

[70] Grant GA, Schuller DJ, Banaszak LJ. A model for the regulation of D-3-phosphoglycerate dehydrogenase, a Vmax-type allosteric enzyme. *Protein science.* 1996;5(1):34-41.

[71] Dey S, Hu Z, Xu XL, Sacchettini JC, Grant GA. D-3-Phosphoglycerate dehydrogenase from Mycobacterium tuberculosis is a link between the *Escherichia coli* and mammalian enzymes. *Journal of Biological Chemistry.* 2005;280(15):14884-91.

[72] Ali V, Hashimoto T, Shigeta Y, Nozaki T. Molecular and biochemical characterization of d-phosphoglycerate dehydrogenase from *Entamoeba histolytica. The FEBS Journal.* 2004;271(13):2670-81.

[73] Mishra V, Ali V, Nozaki T, Bhakuni V. *Entamoeba histolytica* Phosphoserine aminotransferase (EhPSAT): insights into the structure-function relationship. *BMC research notes.* 2010;3(1):52.

INDEX

G

H

I

R

S

T

U

V

W

Y